Fire and Rescue Service

Operational guidance

Incidents in Tunnels and Underground Structures

ESSEX COUNTY COUNCIL LIBRARY

Published by TSO (The Stationery Office) and available from:

Online
www.tsoshop.co.uk

Mail, Telephone, Fax & E-mail
TSO
PO Box 29, Norwich, NR3 1GN
Telephone orders/General enquiries: 0870 600 5522
Fax orders: 0870 600 5533
E-mail: customer.services@tso.co.uk
Textphone 0870 240 3701

TSO@Blackwell and other Accredited Agents

© Crown copyright 2012

Published with the permission of the Department for Communities and Local Government on behalf of Her Majesty's Stationery Office.

Application for reproduction should be made to HMSO, e-mail: psi@nationalarchives.gsi.gov.uk.

ISBN: 9780117541115

Printed in the UK by The Stationery Office Limited

on behalf of the Controller of Her Majesty's Stationery Office

ID2472476 04/12

Contents

SECTION 1
Foreword

Major incidents involving tunnels and underground structures in the United Kingdom are rare. Such incidents place significant demands on local fire and rescue services and often require resources and support from other fire and rescue services and emergency responders. However smaller scale incidents in tunnels and underground structures are more prevalent and these may require a response from any fire and rescue service in England.

The Fire and Rescue Service Operational Guidance – Incidents in Tunnels and Underground Structures provides robust yet flexible guidance that can be adapted to the nature, scale and requirements of the incident.

The Chief Fire and Rescue Adviser is grateful for the assistance in the development in this guidance from a wide range of sources, including the fire and rescue service, Constructors and operators

It is anticipated that this guidance will promote common principles, practices and procedures that will support the fire and rescue service to resolve incidents in this type of structures safely and efficiently.

SECTION 2
Preface

The objective of the Fire and Rescue Service Operational Guidance – Incidents in Tunnels and Underground Structures is to provide a consistency of approach that forms the basis for common operational practices, supporting interoperability between fire and rescue services, other emergency responders, the construction industry and tunnel and underground users. These common principles, practices

and procedures are intended to support the development of safe systems of work on the incident ground and to enhance national resilience.

Operational Guidance issued by the Department of Communities and Local Government promotes and develops good practice within the Fire and Rescue Service and is offered as a current industry standard. It is envisaged that this will help establish high standards of efficiency and safety in the interests of employers, employees and the general public.

The Guidance, which is compiled using the best sources of information known at the date of issue, is intended for use by competent persons. The application of the guidance does not remove the need for appropriate technical and managerial judgement in practical situations with due regard to local circumstances, nor does it confer any immunity or exemption from relevant legal requirements, including by-laws. Those investigating compliance with the law may refer to this guidance as illustrating an industry standard.

It is a matter for each individual fire and rescue service whether to adopt and follow this Operational Guidance. The onus of responsibility for application of guidance lies with the user. Department of Communities and Local Government accept no legal liability or responsibility whatsoever, howsoever arising, for the consequences of the use or misuse of the guidance.

Section 3

Introduction

Purpose

3.1 This operational guidance is set out in the form of a procedural and technical framework. Fire and rescue services should consider it when developing or reviewing their policy and procedures to safely and efficiently resolve emergency incidents involving any aspect of Fire and Rescue Service operations in tunnels and underground structures.

3.2 The term 'tunnels and underground' is a general term that encompasses:

3.3 Transportation and utility tunnels both operational and under construction, caves, potholes, mines, bunkers, underground storage facilities and military installations

3.4 For the purposes of this guidance a tunnel or underground structure is defined as;

'a natural or manmade structure, where all or part is below ground level or covered, where people can resort to for work or pleasure. This includes underpasses, or any associated shafts, but excludes basements'

3.5 Although non Fire and Rescue Service organisations and agencies may use other more specific definitions for their own requirements, the above definition is the most appropriate for fire and rescue services to base their risk assessments and planning assumptions on.

3.6 A fire and rescue service may respond to a wide range of incidents involving tunnels and underground structures that have the potential to cause harm and disruption to firefighters and the community.

3.7 The purpose of this guidance is to assist emergency responders to make safe, risk assessed, efficient and proportionate responses when attending and dealing with operational incidents involving tunnels and underground structures.

3.8 Whilst this guidance may be of use to a number of other agencies, it is designed to provide relevant information in England Fire and Rescue Service relating to planning and operations for incidents occurring in tunnels and underground structures.

Scope

3.9 This guidance covers a wide range of incident types associated with tunnels and underground structures that are likely to be encountered. It is applicable to any event regardless of scale, from incidents, such as a small smouldering fire in an underground structure to large scale multiple vehicle collision in a transport tunnel with significant affect on critical infrastructure, and involving large numbers of Fire and Rescue Service resources and members of the public.

3.10 It focuses on the tactical and technical aspects of tunnels and underground incidents to assist the Fire and Rescue Service with:

- the development of safe systems of work

- inter-operability at large or cross border incidents where more than one Fire and Rescue Service is in attendance
- multi agency working to resolve incidents occurring in tunnels and underground structures.

3.11 This guidance covers the time period from the receipt of the first emergency call to the closure of the incident by the Fire and Rescue Service Incident Commander.

3.12 In addition to detailed tactical and technical information it also outlines the key operational and strategic responsibilities and considerations that need to be taken into account to enable the Fire and Rescue Service to train for, test intervention strategies and plan to ensure effective response to any incident involving tunnels and underground structures.

Structure

3.13 This guidance is based on nationally accepted good practice. It is written as an enabling guide based around risk-critical operational principles rather than a strict set of rules and procedures. This is done to recognise local differences across England and elsewhere in the UK in terms of risk profiles and levels of resource.

3.14 Section 8 contains the bulk of the guidance and is divided into three parts:

- Part A – Pre planning considerations
- Part B – Operational considerations
- Part C – Technical considerations

Part A – Pre-planning
Information supporting Fire and rescue Service personnel in a number of roles when undertaking preparatory work for dealing with incidents that may occur in their service area involving tunnels and underground structures.

This section covers planning considerations at both the strategic level when planning for service wide response options and for those associated with local site specific risks.

Part B – Fire and Rescue Service operational considerations
Guidance to Fire and Rescue Service staff on responding to and resolving typical incidents in tunnels and underground structures. It is structured around six emergency response phases common to all operational incidents.

The procedure detailed in this part of the guidance uses the Incident Command System (ICS) decision making model as its foundation. It is a Generic – Standard Operating Procedure (G-SOP) for dealing with tunnels and underground incidents which fire and rescue service s can adopt or adapt depending on their individual risk assessments and resources.

Each section of the Generic – Standard Operating Procedure details extensive lists divided into:

- Possible actions
- Operational considerations.

It should be stressed that these are **not mandatory procedures.** They are a 'tool box' of operational considerations which will act as an enabling guide when dealing with tunnels and underground incidents.

The Generic – Standard Operating Procedure reflects the hazards and control measures of the national generic risk assessments relevant to tunnels and underground incidents.

Part C – Technical considerations
Contains technical information and operational considerations that may be required by Fire and rescue Service personnel for planning training and operations. It also references more detailed guidance that may be of interest to fire and rescue services.

This part only contains information with an operational connotation and is not intended to be an exhaustive technical reference document.

Section 4

Legal framework

Primary Fire and Rescue Service legislation

4.1 Fire and Rescue Authorities (FRA) need to be aware of the following legislation. It is relevant to command and control at operational incidents and also in the training environment.

4.2 This section does not contain detailed legal advice about the legislation. It is just a summary of the relevant legislation, as applied to FRA firefighting procedures. You should confirm with your legal team on your FRA compliance with this legislation.

4.3 When considering this framework it is essential to recognise that any definitive interpretation of the legal roles and responsibilities imposed by legislation can only be given by a court of law.

4.4 For a full understanding of the responsibilities imposed by the legislation, and by the Fire and Rescue Service National Framework, reference should be made to the relevant legislation or National Framework. It is also recognised that the range of legislation and guidance that could impact on the operational responsibilities of the FRA is extensive and each FRA should seek guidance from their own legal advisors.

4.5 The adoption of the principles set out in this guidance will assist FRA in achieving suitable and sufficient risk assessments and appropriate corresponding risk control measures such as those referred to in this and other similar documents.

Fire and Rescue Services Act 2004

4.6 This is the main Act affecting Fire and Rescue Authorities. Amongst other things, it obliges Fire and Rescue Authorities (in section 7) to secure the provision of the personnel, services and equipment that are necessary efficiently to meet all normal requirements and also to secure the provision of training for such personnel in relation to firefighting.

Fire and Rescue Services (Emergencies) (England) Order 2007

4.7 The Order obliges Fire and Rescue Authorities to make provision for decontaminating people following the release of chemical, biological, radiological, nuclear (CBRN) contaminants (article 2). It also requires them to make provision for freeing people from collapsed structures and non-road transport wreckages (article 3). The Order also obliges Fire and Rescue Authorities to use their specialist chemical, biological, radiological, nuclear or Urban Search and Rescue (USAR) resources outside their own areas to an extent reasonable for dealing with a CBRN or USAR emergency (regulation 5).

Civil Contingencies Act 2004

4.8 Section 2(1) states, among other things, that Fire and Rescue Authorities shall maintain plans for the purpose of ensuring that if an emergency occurs or is likely to occur the Fire and Rescue Authority is able to perform its functions so far as necessary or desirable for the purpose of preventing the emergency, reducing controlling or mitigating its effects or taking other action in connection with it.

The Civil Contingencies Act 2004 (Contingency Planning) Regulations 2005

4.9 Fire and Rescue Authorities must cooperate with each other in connection with the performance of their duties under Section 2(1) of the Civil Contingencies Act 2004. In addition, the Regulations state that Fire and Rescue Authorities may facilitate cooperation by entering into protocols with each other (regulation 7), that Fire and Rescue Authorities may perform duties under section 2(1) jointly with one another and make arrangements with one another for the performance of that duty (regulation 8).

4.10 The Civil Contingencies Act 2004 (Contingency Planning) Regulations 2005 set out clear responsibilities for category 1 and category 2 responders and their need to participate in local resilience forums.

The Regulatory Reform (Fire Safety) Order 2005 (SI 2005/1541)

4.11 Includes Articles relating to the provision and maintenance of equipment and facilities for Fire and Rescue Service personnel.

Primary health and safety at work legislation

Health and Safety at Work etc Act 1974

4.12 This Act applies to all employers in relation to health and safety. It is a wide ranging piece of legislation but in very general terms, imposes the general duty on Fire and Rescue Authorities to ensure, so far as is reasonably practicable, the health, safety and welfare at work of all of their employees (section 2(1)) and to persons not in Fire and Rescue Service employment who may be affected by Fire and Rescue Service activity (section 3(1)). Fire and Rescue Service employees also have a duty to take reasonable care for the health and safety of him/her self and of other persons who may be affected by his/her acts or omissions at work.

Corporate Manslaughter and Corporate Homicide Act 2007

4.13 The Corporate Manslaughter and Corporate Homicide Act was given Royal assent on 26th July 2007, and came into force on 6th April 2008. It is called corporate manslaughter in England, Wales and Northern Ireland, and corporate homicide in Scotland.

4.14 Fire and Rescue Authorities take their obligations under health and safety law seriously and are not likely to be in breach of the new provisions. Nonetheless, Fire and Rescue Authorities should keep their health and safety management systems under review, in particular, the way in which their activities are managed or organised by senior management.

4.15 Any alleged breaches of this act will be investigated by the police. Prosecution decisions will be made by the Crown Prosecution Service (England and Wales), the Crown Office and Procurator Fiscal Service (Scotland) and the Director of Public Prosecutions (Northern Ireland)

Management of Health and Safety at Work Regulations 1999

4.16 These regulations require Fire and Rescue Authorities, among other things, to make suitable and sufficient assessment of the risks to the health and safety of firefighters to which they are exposed while on duty (regulation 3(1)(a)); to implement any preventive and protective measures on the basis of the principles specified in the Regulations (regulation 4); to make arrangements for the effective planning, organisation, control, monitoring and review of the preventive and protective measures (regulation 5) and to provide such health surveillance as is appropriate having regard to the risks to health and safety which are identified by the risk assessment (regulation 6).

4.17 Safety Representatives and Safety Committees Regulations 1977 (as amended) and Codes of Practice provide a legal framework for employers and trade unions to reach agreement on arrangements for health and safety representatives and health and safety committees to operate in their workplace.

4.18 Health and Safety (Consultation with Employees) Regulations 1996 (as amended), sets out the legal framework which will apply if employers have employees who are not covered by representatives appointed by recognised trade unions.

Provision and Use of Work Equipment Regulations 1998

4.19 These regulations require Fire and Rescue Authorities to ensure that work equipment is constructed or adapted as to be suitable for the purpose for which it is used or provided (regulation 4(1)). Fire and Rescue Authorities must have regard to the working conditions and to the risks to the health and safety of firefighters which exist in the premises in which the equipment (including breathing apparatus) is to be used and any additional risk posed by the use of that equipment (regulation 4(2)). The Regulations also contain provisions relating to maintenance, inspection, specific risks, information and instructions and training regarding work equipment.

Personal Protective Equipment at Work Regulations 1992

4.20 These regulations require Fire and Rescue Authorities to ensure that suitable personal protective equipment is provided to firefighters (regulation 4(1)). The Regulations contain provisions in respect of the suitability, compatibility, assessment, maintenance, replacement, storage, information, instruction and training and the use of personal protective equipment.

4.21 Any personal protective equipment purchased by an Fire and Rescue Authority must comply with the personal protective equipment Regulations 2002 and be "CE" marked by the manufacturer to show that it satisfies certain essential safety requirements and, in some cases, has been tested and certified by an approved body.

Control of Substances Hazardous to Health Regulations 2002

4.22 Fire and Rescue Authorities must ensure that the exposure of firefighters to substances hazardous to health is either prevented or, where prevention is not reasonably practicable, adequately controlled (regulation 7(1)). Where it is not reasonably practicable for Fire and Rescue Authorities to prevent the hazardous exposure of firefighters, Fire and Rescue Authorities must, amongst other things, provide firefighters with suitable Respiratory protective Equipment (which must comply with the personal protective equipment Regulations 2002 and other standards set by the Health and Safety Executive.

Dangerous Substances and Explosive Atmospheres Regulations 2002

4.23 Fire and Rescue Authorities are obliged to eliminate or reduce risks to safety from fire, explosion or other events arising from the hazardous properties of a "dangerous substance". Fire and Rescue Authorities are obliged to carry out a suitable and sufficient assessment of the risks to firefighters where a dangerous substance is or may be present (regulation 5). Fire and Rescue Authorities are required to eliminate or reduce risk so far as is reasonably practicable. Where risk is not eliminated, Fire and Rescue Authorities are required; so far as is reasonably practicable and consistent with the risk assessment, to apply measures to control risks and mitigate any detrimental effects (regulation 6(3)). This includes the provision of suitable personal protective equipment (regulation 6(5) (f)).

Confined Spaces Regulations 1997

4.24 A firefighter must NOT enter a confined space to carry out work for any purpose unless it is not reasonably practicable to achieve that purpose without such entry (regulation 4(1)). If entry to a confined space is unavoidable, firefighters must follow a safe system of work (including use of breathing apparatus) (regulation 4(2)) and put in place adequate emergency arrangements before the work starts (regulation 5).

The Work at Height Regulation 2005 (as amended)

4.25 This regulation replaces all of the earlier regulations in relation to working at height. The Work at Height Regulations 2005 consolidates previous legislation on working at height and implements European Council Directive 2001/45/EC concerning minimum safety and health requirements for the use of equipment for work at height (The Temporary Work at Height Directive).

Reporting of Injuries, Diseases and Dangerous Occurrences Regulations 1995

4.26 For the purposes of this section, regulation 3 is particularly relevant because it requires Fire and Rescue Authorities to notify the Health and Safety Executive of any "dangerous occurrences". Some examples of dangerous occurrences as defined in Reporting of Injuries, Diseases and Dangerous Occurrences Regulations 1995 (RIDDOR) relevant to Fire and Rescue Service operations at tunnels and underground incidents include: 'any unintentional incident in which plant or equipment either (a) comes into contact with an un-insulated overhead electric line in which the voltage exceeds 200 volts; or (b) causes an electrical discharge from such an electric line by coming into close proximity to it'.

Specific legislation

4.27 Health and Safety legislation relating to mines is chiefly contained in the Mines and Quarries Act 1954 and in the Health and Safety at Work Act 1974 and regulations made under the two acts. Firefighting and Rescue operations in mines are dealt with in the coal and other mines (fire and rescue) Regulations 1956 (SI 1956/1768).

4.28 The Fire Precautions (Sub-surface Railway Stations) England Regulations 2009.

4.29 The Road Tunnel Safety Regulations 2007.

Further reading

4.30 Operational guidance on the management of risk in the operational environment has been issued in the past. In particular, refer to:

- Fire Service Manual volume 2 (3rd edition) "*Incident Command*"
- Fire and Rescue Operational Assessment Toolkit 2009
- Integrated Risk Management Plan guidance notes
- A guide to Operational Risk Assessment
- Health and Safety Executive's guidance booklet HSG53: Respiratory protective equipment at work: A practical guide
- Striking the balance between operational and health and safety duties in the Fire and Rescue Service.

4.31 The adoption of the principles set out in this guidance will assist Fire and Rescue Authorities in achieving suitable and sufficient risk assessments and appropriate corresponding risk control measures such as those referred to in this and other similar documents.

4.32 The Fire Service College maintain a bibliography of technical guidance to which Fire and Rescue Authorities can refer (Fire Service Manual, Fire Service Circulars, Dear Chief Officer letters, Dear Firemaster Letters, Technical Bulletins, Joint Committee for Design and Development Specifications, British and European Standards, Approved Codes of Practice, Health and Safety Executive guidance). In addition, further technical guidance is available on the Department Communities and Local Government website. However, Fire and Rescue Authorities should also maintain copies of these documents within their own libraries.

Section 5

Strategic role of operational guidance

Strategic perspective

5.1 Strategic managers and fire and rescue services are responsible for ensuring their organisation and staff operate safely when dealing with incidents affecting tunnels and underground structures. Their legal duties and responsibilities are contained in Section 4 of this guidance.

5.2 Fire and rescue services should continually strategically assess the risk, in terms of the foreseeable likelihood and severity, of incidents involving tunnels and underground structures occurring within their areas. This assessment should form part of their integrated risk management plan. The findings will help them to ensure they have appropriate organisation, policy procedures, and resources in place for dealing with incidents of this type.

5.3 How do strategic managers know if they are providing, at least, the minimum level of acceptable service or meeting their 'duty of care'. The following principles may assist Strategic Managers when determining the level of acceptable service and whether they are meeting their duty of care

- operations must be legal and within the requirements of regulations
- actions and decisions should be consistent with voluntary consensus standards, and nationally recommended practices and procedures
- actions and decisions to control a problem should have a technical foundation and be based on an appropriate risk assessment
- actions and decisions must be ethical.

At the incident

5.4 'Response' can be defined as the actions taken to deal with the immediate effects of an emergency. It encompasses the resources and effort to deal not only with the direct effects of the emergency itself (e.g. fighting fires, rescuing individuals) but also the indirect effects (e.g. disruption, media interest). The duration of the response phase will be proportionate to the scale and complexity of the incident.

5.5 The generic key roles of the Fire and Rescue Service at incidents involving tunnels and underground structures are:

- save life and carry out rescues
- fight and prevent fires
- manage hazardous materials and protect the environment
- mitigate the effects of the incident
- ensure the health and safety of fire service personnel, other Category 1 and 2 responders and the general public
- safety management within the inner cordon.

5.6 Separate legislation applies to incidents occurring in mines and quarries, which is covered in more detail elsewhere in this document

5.7 When called to attend a significant incident involving tunnels and underground structures the Fire and Rescue Service has strategic multi-agency responsibilities. These are additional, and in the main complimentary, to the specific fire and rescue functions that the Fire and Rescue Service performs at the scene. The strategic intention is to co-ordinate effective multi-agency activity in order to:

- preserve and protect lives
- mitigate and minimise the impact of an incident
- inform the public and maintain public confidence
- prevent, deter and detect crime
- assist an early return to normality (or as near to it as can be reasonably achieved).

5.8 Other important common strategic objectives flowing from these responsibilities are to:

- participate in judicial, public, technical or other inquiries
- evaluate the response and identify lessons to be learnt
- participate in the restoration and recovery phases of a major incident.

Values

5.9 The Fire and Rescue Service expresses its values and vision of leadership in the form of a simple model. The model has been named Aspire and is fully described in the Fire and Rescue Manual (volume 2 Fire Service Operations – Incident Command). It has at its heart, the core values of the service; which are:

- diversity
- our people
- improvement
- service to the community.

5.10 These values are intrinsic to everything fire and rescue services strive to achieve at an operational incident, where they routinely serve all communities equally and professionally, with the safety and well being of their crews at the forefront of their procedures and reflecting on how well they performed in order to be better next time. It is important that core values are recognised and promoted by all strategic managers and fire and rescue authority members.

5.11 This guidance has been drafted with a view to ensure that equality and diversity issues are considered and developed and has undergone full Equality Impact Assessment (EIA) in line with Priority 1 of the Equality and Diversity Strategy.

Operational guidance review protocols

5.12 This operational guidance will be reviewed for its currency and accuracy three years from date of publication. The Operational Guidance Programme Board will be responsible for commissioning the review and any decision for revision or amendment.

5.13 The Operational Guidance Programme Board may decide that a full or partial review is required within this period.

Section 6

Generic Risk Assessment

6.1 Due to the size and nature of the Fire and Rescue Service and the wide range of activities in which it becomes involved, there is the potential for the risk assessment process to become a time consuming activity. To minimise this and avoid having inconsistencies of approach and outcome, the Department for Communities and Local Government have produced a series of generic risk assessments. These generic risk assessments have been produced as a tool to assist fire and rescue services in drawing up their own assessments to meet the requirements of the Management of Health and Safety at Work Regulations 1999.

6.2 There are occasions when the risks and hazards sited in any of the Generic Risk Assessments may be applicable to incidents in tunnels and underground structures. However there are specific Generic Risk Assessments that Fire and Rescue Services should consider when developing their policy and procedures for dealing with incidents in tunnels and underground structures. They have been used as the foundations of the information and guidance contained in this Operational Guidance.

6.3 Generic Risk Assessments of particular relevance to Tunnels and Underground Structures

- 4.1: Incidents involving transport systems
  – Road http://www.communities.gov.uk/documents/fire/pdf/gra41.pdf
- 4.2: Incidents involving transport systems
  – Rail http://www.communities.gov.uk/documents/fire/pdf/1829947.pdf
- XX Tunnels and underground structures.

6.4 Fire and rescue services should therefore use all relevant, published generic risk assessments as part of their own risk assessment strategy not as an alternative or substitute for it. They are designed to help a fire and rescue service to make a suitable and sufficient assessment of risks as part of the normal planning process. It is suggested that fire and rescue services:

- Check the validity of the information contained in the generic risk assessment against their Fire and Rescue Service's current practices and identify any additional or alternative hazards, risks and control measures
- Evaluate the severity and likelihood of hazards causing harm, and the effectiveness of current controls, for example, operational procedures, training and personal protective equipment etc., by using the Service's methodology
- Consider other regulatory requirements as outlined in Section 4
- Identify additional measures which will be needed to reduce the risk, so far as is reasonably practicable
- Put those additional measures and arrangements in place
- Plan the type, weight and speed of response to be provided to a particular location on the basis of reasonably foreseeable incident scenario. This should be decided on the basis of experience and professional judgement.

6.5 Once a suitable and sufficient assessment of the risks has been made, any additional measures and arrangements put in place have to be reviewed as part of the HSG 65 model

6.6 To ensure the risk assessment remains suitable and sufficient fire and rescue services should review the assessment to take into account, for example the learning outcomes from operational incidents, accidents etc.

6.7 Generic risk assessments provide a guide to the type of information, arrangements and training that should be given to the incident commander, firefighters and any other personnel likely to be affected.

6.8 Full guidance on the generic risk assessments is contained in *Occupational health, safety and welfare: Guidance for fire services: Generic Risk Assessments.*

Section 7

Key principles

Introduction

7.1 This operational guidance offers generic guidance to assist Fire and Rescue Authorities in their preparation for dealing with tunnels and underground incidents as defined in Generic Risk Assessment 2.11 (Operational Incidents in Tunnels and Underground structures). It is essential to consider this guidance and the relevant generic risk assessments in conjunction with local integrated risk management plans and local risk information to develop generic service wide plans, along with site specific variations and adjustments where necessary.

7.2 When planning for incidents involving tunnels and underground structures Fire and Rescue Services should be aware that these can span administrative and governmental boundaries and therefore need to consider the involvement of a range of stakeholders including any Fire and Rescue Service affected.

7.3 During construction and/or refurbishment tunnels and underground structures may present a range of additional complexities for Fire and Rescue Service operations. Fire and Rescue Services should ensure that appropriate intervention plans are drafted and maintained for this period.

7.4 When planning for response to incidents involving tunnels and underground structures fire and rescue services should ensure that relevant aspects of their policies comply with the Confined Space Regulations where appropriate.

7.5 To enhance the effectiveness of the local fire and rescue service and site specific plans the Fire and Rescue Service should ensure that suitable and sufficient training and familiarisation is regularly undertaken to embed understanding of local risks and intervention strategies.

7.6 When developing tactical plans for dealing with tunnels and underground incidents, Incident Commanders will need to use knowledge of pre-planned intervention strategies and take into account all aspects of the circumstances of the incident (e.g.; ventilation, travel distances, available systems such as close circuit television) to ensure that firefighting, and rescue techniques and tactics are appropriate.

7.7 A significant feature of Fire and Rescue Service operations at tunnels and underground incidents is access, egress and evacuation of the public. Incident Commanders should gather sufficient information to facilitate identification of an incident's location and appropriate primary and alternative access and egress points.

7.8 Incidents involving tunnels and underground structures are often by nature linear, with limited access points. This can have a significant effect on the provision of equipment and personnel to scenes of operation. Incident Commanders should therefore carefully consider the affects of the geography and travel distances relating to any incident on logistics, supply chains and crew welfare.

7.9 Due to the complex nature of tunnels and underground incidents, effective liaison at an early stage is essential. Incident Commanders must ensure that timely and appropriate liaison is established with relevant persons and agencies.

7.10 Tunnels and underground incidents can be spread over large areas with command points remote from operations. Incident Commanders should therefore consider the early establishment of effective communications between the key points of the incident management structure.

7.11 Where it is necessary for operational crews to work in tunnels or underground structures, Incident Commanders must ensure that appropriate safety officers are appointed and that they are adequately briefed.

7.12 Incident Commanders should be aware that fires in tunnels and underground structures can produce abnormal fire behaviour including spread and intensity.

7.13 Working conditions at incidents involving tunnels and underground structures can be extremely arduous and resource intensive. Incident Commanders must make an early appraisal of the incident objectives and recognise the limitations of the physical and psychological pressures on any Fire and Rescue Service crews to be committed in achieving those objectives.

7.14 The ability of responders to make an effective intervention is dependent on:

- severity of the incident
- available information and facilities
- any intervention strategies
- availability of resources
- limitations of Fire and Rescue Service equipment.

7.15 In addition to any investigation undertaken by the Fire and Rescue Service, other organisations or agencies may carry out parallel investigations either as statutory duties or as part of service improvement. Therefore fire and rescue services must ensure wherever possible evidence is preserved and records kept of all decisions and actions undertaken during firefighting and rescue operations.

Section 8

Fire Service operations

Part A

Pre-planning considerations

PART A–1
General

8A1.1 Pre-planning at a strategic level to ensure that fire and rescue services develop and maintain an appropriate and proportionate response to tunnel and underground incidents is fundamental to protecting the public, Fire and Rescue Service responders and mitigating the wider impact of any incident.

8A1.2 An appropriate and proportionate response to incidents may include consideration of the following:

- the complexity and relative importance of the infrastructure within its area
- integrated risk management plan response options
- discussion at local resilience fora e.g. threat level
- the hazards associated with the individual infrastructure and the likely severity and impact of any incident
- information received from liaison with industry
- the specific statutory obligations to particular infrastructure, for example mines and tunnels under construction.

8A1.3 For all locations, the features of the infrastructure will influence strategic planners when developing integrated risk management plans, and subsequent operational response. These will include:

- complexity and scale
- the agreed intervention and evacuation strategy
- the local infrastructure's management standards
- the use of the location and likely impact of an incident.

8A1.4 The general duties of the Fire and Rescue Service in responding to emergency incidents are contained within the Fire and Rescue Services Act 2004. Further statutory obligations relating to tunnel incidents are contained in Statutory Instrument 735/2007.

8A1.5 Responsibilities applicable to both Category 1 and Category 2 responders and the direction to participate with Local Resilience Forums are set out within the CCA (Contingency Planning) Regulations 2005.

8A1.6 Many infrastructure managers will have responsibilities under Civil Contingency legislation as 'Category 2' responders to co-operate and share relevant information with Fire and Rescue Service.

8A1.7 The industries involved in managing or constructing tunnels are valued partners, and as such should be incorporated into fire, police and local.

8A1.8 authority emergency response plans. This will be helpful when undertaking planning, training and improving operational response.

8A1.9 Strategic planners will need to consider the contribution to Critical National Infrastructure (CNI) of the United Kingdom that is made by tunnel and underground infrastructure in regard to:

- the national transport networks, with local, national and international dependencies, principally involving road and rail use
- telecommunications and power systems
- water treatment systems
- storage of significant items and use by industries
- the potential for widespread flooding resulting from the inundation of tunnels
- tunnels being put to more than one use, for example a transport tunnel used to carry telecommunications cables, thereby compounding the community impact of a significant incident.

8A1.10 As towns and cities became increasingly complex, major civil works and infrastructure started being carried in tunnels to minimise the disruption and impact of developments. This trend will continue and is likely to increase, involving the following types of infrastructure:

- local, national and international transport systems, involving land and water based transportation including freight and mass passenger transport services and systems
- communication tunnels for cable based systems and high voltage electrical supplies
- large sewerage and water supply systems
- storage or industrial processes
- military or other Crown establishments
- mines.

8A1.11 Strategic planners will need to consider that incidents involving this type of infrastructure can have significant local, national and international impact resulting in:

- widespread disruption
- evacuation and managing large numbers of people
- prolonged and resource intensive operational requirements
- use of specialist equipment, national Fire and Rescue Service resources

- considerable delay due to subsequent investigation and restoration
- loss of public confidence in services and utilities
- loss of critical services, for example civil emergency telephone network, or mass transport infrastructure hub
- significant public and media interest
- subsequent inquiry and public examination of, for example, Fire and Rescue Service planning, intervention and evacuation arrangements, Fire and Rescue Service communications with partners and actions to reduce the impact of the incident and ensure the early restoration of normality.

8A1.12 In order to reduce the impact of an incident in tunnel or underground infrastructure, fire and rescue services should look to develop an agreed integrated operational intervention and evacuation strategy for each location.

PART A-2
Strategic planning considerations and duties

8A2.1 When involved in the development of intervention strategies, for existing or planned refurbishment of tunnels or underground locations, the establishment of appropriate structures and fora, such as the Qualitative Design Review Panel or a Safety Liaison Group should be promoted. This will help to ensure that the impact of a future incident can be minimised. This will inform the development of appropriate integrated and proportionate intervention and evacuation strategies, developed and recorded not only for the period of works, but also for the final infrastructure.

8A2.2 Any intervention strategy, whether for an extensive and complex development, or for a single location (for example a single road or rail tunnel on an otherwise open surface system) should be:

- broad enough to deal with reasonably foreseeable incidents
- simple enough to be easily understood and communicated
- flexible enough to be easily amended to changing circumstances
- provide suitable and sufficient facilities to enable Fire and Rescue Service responders to perform their duties.

8A2.3 The development of an intervention strategy will consider the following aspects of the infrastructure:

- the size, length and method of construction
- level of risk to the public, Fire and Rescue Service and wider potential losses
- the planned evacuation strategy of the public
- the design size fire and assessment of reasonably foreseeable incidents
- the number and location of Fire and Rescue Service intervention points
- any available means of transportation for Fire and Rescue Service crews and resources, and whether this is fit for purpose
- processes undertaken or use of the infrastructure
- facilities provided for Fire and Rescue Service use and measures to protect Fire and Rescue Service personnel and appliances, for example emergency ventilation facilities and water supplies.

8A2.4 During planning meetings, and generally, it would be good practice to ensure, as far as reasonably practicable, that the intervention strategy and operational response to particular locations reflect the policies and procedures that apply to the relevant wider infrastructure. This should mean that, for example:

- rail tunnel incidents should be managed primarily as rail incidents
- road tunnel incidents should be managed primarily as road incidents.

8A2.5 Strategic planners may wish to avoid, if reasonably practicable, the development of different intervention strategies for separate locations, within the same infrastructure. This may be particularly important, and should be promoted, at locations that cross Fire and Rescue Service boundaries and borders. The advantage of this is to:

- reduce the complexity and cost of training for responders
- reduce complexity and cost of training for infrastructure manager's staff
- reduce the likelihood for confusion when responding to incidents
- simplify the information gathering process at incidents for responders
- standardise the method of the infrastructure managers providing information to responders
- achieve consistency in mutual support at cross border incidents.

8A2.6 Fire and rescue services may wish to promote the development of standardised emergency intervention and evacuation strategies for different developments in their area. For example, at tunnels and underground structures different rail infrastructure managers may be encouraged to use essentially similar methods of:

- summoning the Fire and Rescue Service
- liaison with the Fire and Rescue Service and provision of information at incidents
- the operation of any ventilation system
- standardisation of facilities provided to responders.

8A2.7 Effective communications equipment for the Fire and Rescue Service is fundamental to safe systems of work at tunnel or underground incidents. Strategic planners will need to satisfy themselves that for any intervention strategy, appropriate, proportionate and effective equipment is in place. Any communications equipment provided should be robust, suitable and sufficient enough to support the agreed intervention strategy and fire and rescue service safe systems of work.

8A2.8 The provision of infrastructure equipment will usually be supplied and maintained by the infrastructure owners or occupiers.

8A2.9 Any agreement with the owners or occupiers as to the arrangements necessary to support emergency responders, including silver liaison arrangements, should be recorded formally. It should be emphasised that during an emergency response the Fire and Rescue Service will not wish to be involved in identifying areas of administrative demarcation. It would be the expectation of responders that any support provided to Fire Silver would normally be seamless and invisible, involving a single point of contact, normally at the incident.

8A2.10 The scale of the infrastructure manager's response should be proportionate to the size, complexity and vulnerability of the underground structure and persons who may be at risk, taking into account any facilities provided, as well as the likely impact of any incident.

- The Fire and Rescue Service may wish to express the expectation that the person providing the single point of contact at Silver will be expected to:
- have the authority, knowledge, training and background to make and implement decisions at Silver level:
- carry out liaison with and represent any other occupiers or their agents or contractors, forming a single point of contact for emergency responders.

8A2.11 Planners will recognise that older infrastructure may not have the desired standard of intervention facilities. Strategic planners should measure the capabilities of responders and equipment against the risks associated with the infrastructure and the impact of its loss; developing plans, procedure and equipment if appropriate.

8A2.12 Arrangements for routine liaison with infrastructure managers may be established to ensure liaison at Bronze, Silver and Gold levels takes place as appropriate.

8A2.13 When developing an intervention strategy, planners will be cognisant of any existing 'major incident procedures' and seek to ensure that these are complimentary.

8A2.14 When developing strategies for any emergency response to a mine incident, the Fire and Rescue Service will make particular reference to the separate statutory duties that apply to those locations.

8A2.15 Strategic planners should be as mindful of the intervention and evacuation arrangements required during the construction or development stages of infrastructure, as for the final commissioned infrastructure. For tunnels under construction BS 6164 should form the basis of planning an operational response.

8A2.16 During the planning process it should be made clear that the infrastructure manager's emergency plans for evacuation of staff and public must not rely upon any Fire and Rescue Service response. It should be emphasised that the emergency plan should be sufficiently robust and flexible to effectively manage the safe evacuation of people without being dependant on Fire and Rescue Service resources.

PART A–3
Multi-level planning

8A3.1 For existing infrastructure the planning for the Fire and Rescue Service response can be undertaken as part of the Fire and Rescue Service usual business, reflected integrated risk management plans in liaison with local industries, services and strategic bodies.

8A3.2 Construction or significant redevelopment work would normally require more detailed and on-going commitment through planning and construction to the commissioning stage. This may take place at three principal levels. An example of good practice is shown below.

8A3.3 If an incident on the infrastructure attracts a response from more than one Fire and Rescue Service, an agreement should be made at a strategic level as to which service would take command, and the support arrangements to be implemented.

8A3.4 During the planning and design stages of a tunnel or underground project the use of Qualitative Design Review meetings detailed in BS 7974 may be helpful in determining the detailed equipment and facility requirements necessary to satisfy the Fire and Rescue Service needs in terms of operational intervention and evacuation strategies.

Level	Description	Attendees
Strategic Gold (Level 1)	Safety Liaison Group This is a planning committee comprising of strategic level planning representatives from Category 1 and 2 responders, enforcers, client and developer's representatives. This group would deal with, for example, the development of intervention and evacuation strategy. Agreeing design principles for the strategy and public safety issues for the final commissioned system. It may also deal with Fire and Rescue Service issues during the construction stage that cannot be resolved at Silver/ Bronze level.	Attendees include: Appropriate police, fire and rescue services and ambulance representatives, relevant statutory enforcers such as Health and Safety Executive, Her Majesty's Inspector of Railways from the Office of the Rail Regulation (ORR), interfacing industries (Network Rail), local authority emergency planners, strategic bodies
Tactical Silver (Level 2)	'System-wide' liaison between developers and the Fire and Rescue Service to: • review recent operational incidents and identify learning opportunities • consider the development's progress and amend construction stage intervention and evacuation strategies accordingly. • share relevant organisational policy information • discuss Fire and Rescue Service observations.	Fire and Rescue Service operational policy and regulatory Fire Safety representatives Developers
Operational Bronze (Level 3)	Typically this will include: • 7(2)(d) visits for Information gathering and local operational pre-planning • training and liaison • identification of site specific issues.	Local Fire and Rescue Service responders and local managers

PART A-4
Underpinning strategic knowledge

8A4.1 The day to day management of tunnels and underground structures can involve various stakeholders, complex and hazardous environments, machinery and processes. To inform and support strategic planning it is essential that Fire and Rescue Service personnel tasked with developing emergency response plans should have some underpinning knowledge with regard to:

- the statutory duties and limitations placed upon the Fire and Rescue Service
- the statutory duties of the owner/occupier/contractor
- the unique challenges presented to emergency responders in tunnels and underground locations. This is particularly the case for incidents involving hazardous materials or complex rescues
- the level of facilities that may reasonably be provided for fire and rescue purposes
- the responsibility to manage commercial or security sensitive information appropriately
- balancing the need for operational access against the desired level of security, to ensure effective response times
- the geographical and administrative boundaries of different tunnel or underground systems
- the geographical boundaries of responsibility ensuring pre-planning with appropriate representatives from strategic and local partners
- the location of any control rooms for different infrastructure, both on-site and remote, controlled by the infrastructure manager
- the hazards involved and control measures necessary
- different emergency procedures and any staff responsibilities in different infrastructure, and that these can be involved at the same incident
- differing responsibilities and authority of workers on site
- effective arrangements for establishing meaningful on-site Silver liaison, and interim alternatives
- the numbers of people likely to be involved in an incident
- Fire and Rescue Service intervention tactics.

8A4.2 Liaison is necessary in the development of a strategic operational response strategy and with developing and maintaining plans, procedures and maintaining appropriate contacts. The objective of liaison may be to achieve:

- combined intervention and evacuation strategies for complex tunnel infrastructure or particular industries or uses
- agreed operational response to shared infrastructure with neighbouring fire and rescue services, using an agreed intervention strategy
- determine Information gathering methods for operational personnel in local planning and at the scene of an incident
- methods to exchange information for improving emergency response with partners, including emergency services and other agencies based on operational experience.

8A4.3 The diagram below shows the interface between infrastructure managers' responsibilities and Fire and Rescue Service responsibilities with respect to intervention concepts fin tunnels.

8A4.4 Any strategy developed will normally look to provide safe systems of work to allow Fire and Rescue Service operations to commence before the arrival of an industry Silver representative. Exemption from this will normally be because the infrastructure is very high risk and not normally occupied by the public, for example high voltage national grid tunnels.

8A4.5 A range of activities may be undertaken by a Fire and Rescue Service to examine performance of plans following exercises or incidents with a significant impact. This may include the development of systems and processes to analyse and review the performance of plans, and/or liaison and exercise at all levels with:

- managers of the appropriate level
- other emergency services
- other statutory agencies.

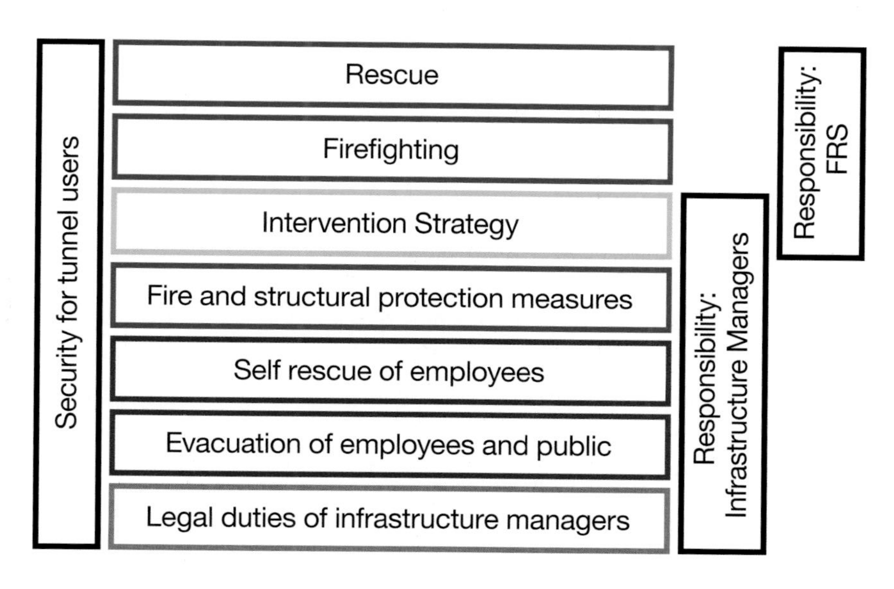

PART A–5
Future developments

8A5.1 As part of consultation on future developments, it is important for the Fire and Rescue Service to actively participate during the development of emergency planning assumptions and of an appropriate intervention strategy. Further, it is necessary to test the assumptions and determine appropriate facilities for evacuation and operational intervention through a system, for example a Qualitative Design Review process or Safety Liaison Group. This will be beneficial in providing assurance that, for example, the necessary extinguishing facilities are provided for dealing with the design size fire. This will ensure that the local fire and rescue service's and infrastructure managers' expectations are realistic and reasonable. Additionally it is important to examine the consequences of Fire and Rescue Service intervention on the tunnels own evacuation procedures.

8A5.2 When a new infrastructure development is planned it would be prudent for Fire and Rescue Services to identify similar developments elsewhere in order to compare the strategy and facilities being proposed against similar examples and examine any lessons learnt.

8A5.3 To ensure a consistent approach to Fire and Rescue Service operations, it is recommended that intervention strategies are formally agreed with managers, owners or operators. This is particularly appropriate in areas involving complex infrastructure or used for mass transit transport systems. Ideally infrastructure being put to similar use, for example road tunnels, should be encouraged to achieve a commonality of approach in terms of the strategic transport bodies actions and response to any road tunnel incident. This is particularly important for features such as:

- intervention systems
- facilities provided for Fire and Rescue Service response
- communications between Infrastructure Managers and Fire and Rescue Service
- liaison with Responsible Person at Silver.

8A5.4 In order to ensure Fire and Rescue Service responders can effectively and efficiently respond to a significant incident in tunnel or underground infrastructure training on the appropriate command and tactics to be employed should be provided to those likely to respond. This is particularly appropriate to incidents involving:

- fires
- structural collapse
- terrorist acts of aggression
- incidents with an environmental impact
- inundation and flooding.

PART A–6

Local planning responsibilities

8A6.1 When developing local plans it is important to consider that incidents underground or in tunnels present special hazards due, amongst other things, to the following characteristics, which many underground incidents have in common:

- a possible absence of adequate means of ventilating the heat, smoke or toxic products of an incident
- potential difficulty in accessing and way-finding
- the potential for rapid escalation
- potential difficulty in appraisal of any fire condition, or of the whereabouts of the fire itself, without risk to firefighters entering dangerous places
- difficulty in communicating between people on the surface and below ground
- difficulty in effectively applying any extinguishing media
- unusual congestion and restriction of movement in the underground space.

8A6.2 This section is intended to inform and advise at Silver and Bronze levels on the development of local plans. Some examples of where local plans would be required range from, the development of an intervention strategy for a new tunnel development within an Fire and Rescue Service area through to a local fire station's site specific pre-determined on arrival tactics.

8A6.3 The development of local plans should reflect Fire and Rescue Service policies; procedures and local risk assessments developed at the strategic level (see Section 5). Local plans should also reflect the guidance within this manual.

8A6.4 Suitable arrangements should be put in place to gather relevant information to facilitate the development of local plans for all of the underground risks and types of incidents that local fire and rescue services may be called to attend.

8A6.5 Plans should also be developed for any significant temporary local works and/or variations to existing plans affecting any underground risks, with consideration given to the affect of those changes on weight of attack, tactics and previous agreements.

PART A–7
Local planning liaison

General

8A7.1 Liaison with various stakeholders is essential to ensure that the necessary information is secured to inform plans for adequate, timely and effective response and to create a safe system of work when planning for and attending incidents.

Industry liaison

8A7.2 Fire and rescue services should be aware that there are a wide range of separate industries that can be involved in planning for incidents occurring in tunnels and underground structures. Industry specific information is available from a variety of national and local sources, some examples include:

- Department for Transport Rail Group
- Network Rail
- Association of Train Operating Companies
- Freight Transport Association
- London Underground
- Highways Agency
- Office of the Rail Regulator Her Majesty's Inspectors of Railways
- Road Tunnel Operators Association
- Health and Safety Executive
- Local resilience groups
- Water and sewage companies
- Her Majesty Inspector of Mines
- Mines Rescue Service Ltd
- Defence Fire Risk Management Organisation (DFRMO)
- Cave rescue organisations.

Liaison

8A7.3 Many tunnels and underground structures will span organisational and administrative boundaries and are likely to be complex and resource intensive, often involving protracted operations. Therefore incidents of this type may, require responses from a number of emergency services, specialist teams and neighbouring fire and rescue services. This should be taken into account and as a result close liaison between Regional partners is essential when pre-planning.

PART A–8

Local planning information

General

8A8.1 Following relevant research, fire and rescue services should ensure that detailed local plans are prepared to include some or all of the following information. It is imperative that this information is made available to all relevant personnel prior to and upon arrival at incidents. This will ensure that work activity is planned, supervised and carried out safely.

Local knowledge

8A8.2 Under Section 7.2(d) of the Fire and Rescue Services Act 2004, fire and rescue services should make arrangements for operational personnel who are likely to attend incidents within tunnels or underground structures to familiarise themselves with the risk and the likely hazards.

8A8.3 Particular attention should be given to the following:

- depth and dimension of tunnel or underground structure
- access and egress points
- rendezvous points
- significant hazards such as: power supplies, deep water, hazardous substances, confined space
- intervention points/emergency response locations
- infrastructure control rooms
- potential bridgeheads
- ventilation
- fixed installations.

Depth and dimension of tunnel or structure

8A8.4 Some underground structures or parts of will be beyond the capability of Fire and Rescue Service personnel and equipment. The extent to which Fire and Rescue Service safe systems of work will apply should be ascertained during pre-planning. Where operational limitations are identified, these should be formally notified and agreed in writing with the relevant infrastructure manager.

Access

8A8.5 All practical and reasonable areas of access, on to the underground infrastructure which may include:

- stations (both surface and sub surface)
- tunnels
- intervention points/emergency response locations
- access shafts
- bridges
- culverts
- gates and hard standing for appliances
- appropriate maintenance access points.

Rendezvous points

8A8.6 When determining the most suitable position for RENDEZVOUS POINTS, consideration must be given to:

- crew safety and welfare
- access and hard standing for appliances
- effective communications
- plans boxes
- water supplies
- fixed Installations.

Hazards to firefighters

8A8.7 In general the number of hazards facing firefighters and the likelihood of the associated risks occurring will vary in line with complexity in the underground infrastructure and its geographical location. This information should be considered in conjunction with the national generic risk assessments and technical information within this guidance.

Intervention points/emergency response locations

8A8.8 These are locations that can be used for means of access for an emergency response. Emergency Response Locations (ERL) will also provide integrated facilities for Fire and rescue Service intervention and managed evacuation by the relevant infrastructure manager They may also incorporate evacuation facilities for members of the public. They can vary greatly from basic access stairs to complex purpose built structures. Crews should be aware of the following features:

- their location
- rendezvous points
- access arrangements
- plans
- water supplies
- communication facilities.

Infrastructure control rooms

8A8.9 An understanding of the facilities afforded by infrastructure control rooms will assist in determining the means by which an incident can be managed, these may include:

- location
- alternative access/egress
- close circuit television
- public address systems
- ventilation systems.

Ventilation systems

8.10 Ventilation systems may assist in the control of the fire/accident environment. Crews should be aware of the type, location, and operation of the control systems. Types of system are described elsewhere in this guidance.

Note:
Ventilation systems may assist in the control of the fire/accident environment. Crews should be aware of the type, location, and operation of the control systems. At incidents involving fire or hazardous materials ventilation systems should not be turned off or re-configured until a risk assessment has been made and the full consequences of these actions to the public, firefighters and any fire development are known. Further information on ventilation systems are described elsewhere in this guidance.

Fixed installations

8A8.11 Fixed installations to assist firefighting operations are provided in some underground structures. The location, use, and implications of their operation should be known and understood. Fixed installations available may include:

- automatic fire detection systems
- sprinkler systems
- inert gas systems
- 110 volts electrical supplies for Fire and Rescue Service use
- fire mains/hydrants
- communications systems.

PART A–9

Availability of local plans/training/review

Availability of plans

8A9.1 Local plans should be readily available in appropriate formats to support the needs of first response crews, Incident Commander's, command support and elsewhere as required by local fire and rescue service arrangements and the National Incident Command System. Where appropriate, consideration should also be given to sharing plans with other agencies and organisations.

Training and exercising

8A9.2 Effective and realistic training programmes will help prepare Fire and Rescue Service personnel for the variety of challenges, which may be encountered when dealing with underground incidents. The mutual benefits of joint training and exercising include:

- greater inter service understanding
- confidence and awareness of relevant roles and functions
- local familiarisation with key personnel and features of infrastructure.

8A9.3 Local fire and rescue services will wish to prepare and develop fire crews awareness of on arrival, intervention tactics and review of operations against the National Occupational Standards on Emergency Fire Service Management (EFSM), through:

- on site testing of fire and rescue service aspects of plan
- full scale multi-agency exercises
- 7(2)(d) and 9(3)(d) visits, internal lectures and presentations for Fire and Rescue Service responders.

Review

8A9.4 Many tunnels and underground structures are subject to significant on-going change and modification. Therefore fire and rescue services should ensure that local plans are regularly reviewed and updated. This may be either periodically or at key milestones in the case of refurbishment/construction projects.

8A9.5 Following any significant incident/accident involving tunnels or underground structures, a full and robust review of all Fire and Rescue Service policies, standard operating procedures (SOPs) and memorandums of understanding (MoUs) should be undertaken. The outcome of these reviews should result in all Fire and Rescue Service policies, standard operating procedures and memorandums of understanding being amended where required.

Part B

Operational considerations – Generic Standard Operating Procedure

Introduction to Generic – Standard Operating Procedure (G-SOP)

It is useful to see the emergency incident response phases in the context of the typical stages of an incident as referred to in *Volume 2 Fire Service Operations Incident Command Operation Guidance* and the Fire Service Guide – *Dynamic Management of Risk at Operational Incidents*, this is shown below:

Stages of an Incident (Dynamic Management of Risk)	ICS Decision Making Model Links	Generic Standard Operating Procedure Response Phases
		1. Mobilising and En-route
Initial Stage	• Incident information • Resource information • Hazard and safety and information	2. Arriving and Gathering Information
Development Stage	• Think • Prioritise objectives • Plan	3. Planning the Tactical Plan
	• Communicate • Control	4. Implementing the Tactical Plan
	• Evaluate the outcome	5. Evaluating the Tactical Plan
Closing Stage		6. Closing the Incident

The Generic Standard Operating Procedure has been derived by breaking down an incident into six clearly identified phases which have been taken directly from the decision making model.

The purpose of this section is to cover possible actions that may need to be undertaken at each of the six stages of the incident and then offer some possible considerations that the incident commander and other fire and rescue service personnel may find useful in tackling the challenges and tasks that they are faced with.

This Generic Standard Operating Procedure is not intended to cover every eventuality however it is a comprehensive document that can be used by planning teams, who need to write standard operating procedures, and responding personnel alike.

Further detailed and technical information on specific hazards related to incidents in tunnels and underground structures are covered in Section 8 part C of this operational guidance.

The decision making model comprises of two major components. These are the deciding and acting stages.

DECIDING	ACTING

In seeking to resolve incidents in tunnels and underground structures an Incident Commander will use their knowledge and experience to identify the objectives to be achieved and formulate an appropriate tactical plan of action.

Emergency incident response phases

1	Mobilising and en-route
2	Arriving and gathering information
3	Planning the response
4	Implementing the response
5	Evaluating the response
6	Closing the incident

PART B–1

Mobilising and en-route

Possible actions

Phase 1 - Actions Mobilising and En-route	
1.1	Initial call handling
1.2	Assess the level and scale of the incident
1.3	Mobilise appropriate resources to the incident, marshalling areas and/or predetermined rendezvous points (RVPs)
1.4	Access incident specific information en-route
1.5	Notify relevant agencies

Considerations

1.1 Initial Call Handling

8B1.1 As with any incident the handling of the initial call is of critical importance to ensure that the correct predetermined attendance (PDA) is mobilised. In handling the call the mobilising centre operator will need to gather as much information from the caller as possible. If there is any doubt as to the size and scale of the incident, the predetermined attendance should be scaled up rather than down.

1.2 Assess the level and scale of the incident

8B1.2 Irrespective of the tunnel design or management systems employed the potential for emergency incidents cannot be totally removed. Fire and Rescue Service control staff and systems should be able to quickly and accurately match the information given by callers to site specific pre-determined attendances, this combined with the local Fire and Rescue Service mobilising policy will determine the weight and scale of attendance which should reflect the foreseeable risks posed by the incident type and the site/location of the incident

1.3 Mobilise appropriate resources

8B1.3 Fire and Rescue Service Controls should utilise any site specific plans to enhance mobilising information to crews and other responding agencies where appropriate. This is particularly relevant when mobilising to locations where Fire and Rescue Service responders do not have sole responsibility for rescue or firefighting operations, or where the location may be complex, sensitive or particularly high risk.

8B1.4 Specific risks will attract a range of different local mobilising solutions, these will normally be determined in the planning stage and may include:

- the mobilisation of specific resources
- the mobilisation of various responders
- variations in weight of attack
- attendance to specific locations
- attendance to more than one entry point
- requesting the support of other agencies.

1.4 Access Incident specific information en-route

8B1.5 A key aspect for dealing with incidents in this type of infrastructure is securing effective access to the scene that will support safe systems of work.

8B1.6 Fire and Rescue Service appliance commanders should access site/incident pre planning information from on board systems whilst en route to identify: rendezvous points, communication facilities, pre determined on arrival tactics, types of ventilation available, points for initial information gathering on arrival.

8B1.7 Known underground locations should where possible have pre determined intervention strategies; early identification of alternative access points is of benefit when tactical planning and allows for incidents to be tackled more effectively.

8B1.8 Mobilised crews should share and discuss information and begin to think about the hazards they are likely to face, making early considerations of corresponding control measures.

1.5 Notify relevant agencies

8B1.9 Fire and Rescue Service Controls should maintain contact details of responsible organisations for underground risks in their area. This may include various transport infrastructure managers, utility companies or mine owners.

8B1.10 Fire and Rescue Service Controls should be aware of intervention strategies where it has been agreed that certain agencies or organisations may attend to assist with on site liaison or provide information, and pass this information to mobilised crews.

8B1.11 The relevant organisation may be aware of incidents occurring on their infrastructure. However it is good practice for Fire and Rescue Service Controls to inform the relevant organisation of any incidents being attended by their Fire and Rescue Service.

8B1.12 Additional notification procedures may be appropriate for known locations that may be:

- subject to additional legislative requirements (for example mines or infrastructure required to provide their own rescue or firefighting arrangements), or
- sensitive locations (for example military or defence establishments).

8B1.13 The importance of the information gathered by Fire and Rescue Service Control at the initial stage cannot be overstated in establishing the Fire and Rescue Service response when mobilising the appropriate resources e.g.:

- location
- incident type
- status of the significant hazards (for example, high voltage electricity, moving vehicles).

8B1.14 This will be particularly useful where Fire and Rescue Service resources may be travelling long distances, in remote locations or where the call is to complex areas of the infrastructure.

PART B–2

Phase 2 – Arriving and Gathering Information

Possible actions

Phase 2 – Actions Arriving and gathering information	• Incident information • Resource information • Hazard and Safety and information
2.1	Confirm location
2.2	Confirm incident type
2.3	Identify access routes
2.4	Identify features and facilities of the infrastructure involved
2.5	Confirm use of infrastructure
2.6	Gather information and liaise with persons on scene
2.7	Use local knowledge
2.8	Identify available resources
2.9	Identify risks and hazards

Considerations

2.1 Confirm location

8B2.1 The seriousness of the incident in a tunnel or underground premises or structure may not be immediately apparent. There is also the potential for a sudden unexpected development of a fire to occur. Crews should be committed to investigate the type and severity of the incident, after being made aware of the potential hazards, control measures, and being appropriately equipped.

8B2.2 Upon arrival make contact with the responsible person or their representative, if available.

8B2.3 In addition to the identification of the most appropriate access point the precise location of the incident within the infrastructure must be sought in order to:

- identify the likely travel distances and working duration of crews
- identify or anticipate any blockages that may affect access, or alternative access routes

- factor the likely fatiguing effects of extended travel distances and carrying equipment including return journey's into tactical planning.
- determine the appropriate use of any vehicle that may be available, and suitable for carrying personnel or equipment
- the likely resource requirements to initiate and maintain an effective intervention
- estimate the time required to establish an effective intervention and consider the likely development of the incident during that time.
- identify appropriate bridgeheads or equipment staging areas
- consider the complexity of the locations (even connected twin tunnels can have a disorientating effect, if there are no 'way finding' or 'location indicators' available)
- the hazards that exist en-route to and at the incident scene
- the design and topography of the infrastructure considering:
  - the position of any ventilation outlets, where the products of the incident may affect those on the surface or remote from the incident
  - the direction of any mechanical forced ventilation providing safe areas for members of the public and operational bridgeheads
  - any 'piston' effect or other non-controlled air movement
  - the gradient of any passageway allowing run off, liquid contamination or flowing fuel fire to spread, or the potential for inclined surface 'trench effect'.
  - the method of containing run off or contaminated liquids, and the environmental impact
  - the stability of the structure and it's affect on the surface, or risk of inundation of the infrastructure
  - identify the possible spread of flood water and where applicable it's predicted effects on the wider community, if inundation occurs.

8B2.4 Incident Commanders may use a variety of information sources to inform this such as:

- local pre-planning information
- mobilising information
- information from occupiers, owners or infrastructure managers
- local risk information
- on site information provided for Fire and Rescue Service use
- on-site or industry emergency crews.

8B2.5 On arrival Incident Commanders and crews may use a number of additional means to identify the precise location such as:

- on site plans or plan boxes
- way finder marking
- location signage
- emergency telephone references/numbers
- cross passage door numbers
- electrical substation name plate
- features of the locations cross passage, shaft, and equipment
- significant features.

8B2.6 Once the precise location is confirmed Incident Commanders must ensure that an informative message is sent to Fire and Rescue Service Control.

8B2.7 In addition to the above features crews, once committed, should be aware of the uses and limitations of breathing apparatus guidelines and hoselines to assist in way finding.

2.2 Confirm incident type

8B2.8 There are various types of incident which may occur in tunnel and underground infrastructure:

- fire on infrastructure
- fire on vehicles
- collisions, including road traffic collisions or derailment
- person(s) trapped on or in vehicles or machinery
- persons lost or fallen into underground structure
- flooding/inundation
- hazardous material (Hazmat)
- explosions
- collapse
- aggressive acts involving any of the above.

8B2.9 Some infrastructures will contain large numbers of people, unfamiliar with their surroundings or emergency procedures. The responsibility for their evacuation in an emergency rests with the infrastructure managers.

8B2.10 Fire and rescue services will undertake rescues of staff and/or members of the public where they are in imminent danger, however it is not the responsibility of the Fire and Rescue Service to ease a tunnel management's operational difficulty.

8B2.11 Incident Commanders should attempt to identify the progress and success of managed evacuation. If it appears that, for whatever reason, the management is inadequate and people are, or may be imminently, exposed to harm then the situation becomes a rescue and the Incident Commander will take appropriate action.

2.3 Identify access routes

8B2.12 It is always preferable for Fire and Rescue Service crews to gain access to the infrastructure by means designed for public or fire and rescue service access purposes. This would normally be via:

- emergency response locations/intervention points (ERL/IP)
- tunnel portals
- public or worker access points (entrances, stations, stairwells, maintenance shafts).

8B2.13 Some entrances to tunnels or underground structures will not be appropriate until specific control measures are implemented following a risk assessment. This may be because of the type of machinery or process underway, examples are:

- shafts solely occupied by fan blades or large machinery without provision for local isolation
- no facilities for reaching the scene, without improvisation
- no facilities for transporting equipment between floors or levels
- direct access to high speed vehicle movement (rail or road), without warning, lighting or conspicuous indication of the risk.

8B2.14 The Incident Commander will assess the urgency of the situation, limitations of equipment and personnel and the likely development when determining the most appropriate method of accessing the infrastructure as part of any risk assessment.

8B2.15 Fire and Rescue Service personnel must not move from an area intended for normal public use (e.g. station platforms, open public walkway) to an area where they may be exposed to a disproportionate hazard from, for example, rail or road vehicles, machinery, high voltage electricity or infrastructure, without first implementing appropriate control measures. Any signage provided should be considered as part of any risk assessment.

2.4 Identify features and facilities of the infrastructure involved

8B2.16 On arrival Fire and Rescue Service crews should use local knowledge of the infrastructure to which they are responding in order to assist the Incident Commanders in determining:

- who is responsible for the management of the infrastructure (this may involve more than one infrastructure manager or occupiers)
- what nature of hazards is likely to be encountered (for example, road/rail vehicle movements, high voltage electricity, falls from height, irrespirable atmospheres)
- features of the infrastructure that may represent additional risks to firefighters including:
  - road vehicles
  - rail vehicles
  - rapid fire spread
  - limited fire protection or smoke/fume control
  - hazardous materials in storage or transport
  - electrical hazards
  - confined space
  - security features (blast doors, self locking doors, entry control systems and crews moving into areas under the control of other occupiers)
  - electrical systems including those outside the control of the infrastructure manager such as National Grid cables in transport tunnels
  - construction or mining features including falls from height (down a shaft), plant, construction or mining vehicle movements
  - fumes or gasses, either naturally occurring or as a consequence of vehicles, machinery or processes.
- Facilities of the infrastructure that may be provided for Fire and Rescue Service use including:
  - plans or plan boxes
  - Intervention point/evacuation point
  - ventilation systems
  - bridgeheads/firefighting lobbies
  - fire rated vision panels/spy holes
  - Fire and Rescue Service walkways
  - firefighting mains

- hard standing
- communications systems
- on-site information
- emergency lighting
- electrical supplies for Fire and Rescue Service use
- gas monitoring/pollution/untenable atmosphere indicators
- warning signage/operator signage for responder's use
- fixed firefighting and suppression systems.

2.5 Confirm use of infrastructure

8B2.17 Incident Commanders should gather relevant information regarding the specific use of any relevant infrastructure in terms of:

- whether location is used by the public
- if other agencies or organisations have responsibility for emergency management or intervention, and what actions they have taken or are about to take
- use of mass transit and the associated transport of dangerous goods
- any evacuation information/strategies implemented
- any information regarding mobility impaired persons and progress on infrastructure manager's duty to evacuate
- type and quantity of freight carried
- information on hazardous materials or processes stored or in use.

2.6 Gather information and liaise with persons on scene

8B2.18 Incidents in tunnel or underground structures can involve more than one intervention point which can span Fire and Rescue Service or national boundaries. Other emergency services, agencies and commercial undertakings may also be on site and able to provide information to Fire and Rescue Service responders. Gathering and providing relevant timely information to and from interested parties is critical to successful outcomes.

8B2.19 Incident Commanders should identify those representatives on site and others that may provide quality information, establishing appropriate liaison structures, up to and including locally agreed arrangements for dealing with major incidents.

8B2.20 The following may be involved in exchanging operationally relevant information, at some point of an incident:

- Local police service
- British Transport Police
- Infrastructure manager's representative
- Occupiers representative
- Transport operating companies or strategic transport Organisations (Highways Agency)
- Environment Agency
- HMIM/mines rescue service/mine owners
- Local authorities
- Primary care trusts
- Utility companies
- Statutory investigators (Health and Safety Executive or Rail Accident Investigation Branch).

8B2.21 The use of structured silver meetings early in the incident will assist in sharing important or critical information.

8B2.22 Other information may be gathered from:

- plans boxes
- survivors
- witnesses
- staff
- incident indicators including:
  - volume, colour and energy in smoke issuing
  - symptoms and behaviour of casualties.

2.7 Use local knowledge

8B2.23 Fire and Rescue Service crews and Incident Commanders should use knowledge of previously agreed intervention strategies, these strategies being either site specific or relevant to the wider type of infrastructure. Additionally, experience of exercises or familiarisation visits will inform operational strategies and tactics.

8B2.24 Care should be taken to ensure that the location of any low level air inlets is identified to prevent fumes or contaminants entering infrastructure.

2.8 Identify available resources

FIRE AND RESCUE SERVICE RESOURCES

8B2.25 The Fire and Rescue Service has a range of specialist and non-specialist equipment and trained personnel to assist in dealing with various aspects of incidents in tunnels and underground structures, examples of which are:

a. detection identification monitoring equipment

b. enhanced command support

c. hazmat officers

d. high volume pumps

e. incident response units

f. inter-agency liaison officers

g. press liaison officers

h. thermal image cameras

i. short circuiting devices

j. specialist rail vehicles

k. urban search and rescue dog teams

l. urban search and rescue modules and teams.

8B2.26 Incident Commanders will need to gather relevant information about Fire and Rescue Service resources in attendance and en route and assess whether additional local or national resources are required.

NON FIRE AND RESCUE SERVICE RESOURCES

8B2.27 Infrastructure managers may have access to equipment and personnel that may be available to assist fire and rescue services. This can be identified through local liaison and requested via agreed pre planned procedures. This may include:

- telecommunications equipment
- lighting
- air shelter
- bulk drinking water supplies
- video cameras and playback facilities
- Air Wave fitted to command vehicle's and helicopter
- heavy lifting equipment and cranes
- specialist rescue teams or facilities for 'confined space' areas or caves and mines

- various specialist teams to assist in:
  - lifting and moving rail and road vehicles
  - body recovery
  - humanitarian support
  - customer care and survival reception centres.

8B2.28 In some instances, protected 'Control Rooms' are present within the infrastructure providing facilities to assist the management of an incident. These may include facilities for Fire and Rescue Service personnel to monitor progress of crews and the safe evacuation of the public. Alternatively, for wide spread or complex infrastructure a central control room may be provided. Both of these locations may provide a range of facilities for fire and rescue service use including:

- close circuit television
- Fire and Rescue Service radio communications including main scheme and incident ground
- plans
- ventilation controls
- REFUGE communications
- open door indicators
- public address systems
- wipe boards
- faxes
- telephones
- desks and seats
- sector competent experts.

8B2.29 In most circumstances emergency ventilation and evacuation procedures will be implemented before the arrival of the Fire and Rescue Service or other responder either automatically or by the infrastructure manager.

8B2.30 Once ventilation systems are activated, it will not normally be necessary for the Fire and Rescue Service or others to alter the direction or flow of a ventilation system. However, some ventilation systems can be adjusted to increase/decrease flow to certain areas or change direction.

8B2.31 This could be used to positive advantage for fire and rescue service operations, however the effects of any changes could have disastrous consequences for people moving away from smoke or contamination to a perceived place of relative safety or escape. Therefore adjustments and alterations after the arrival of responders should only be made on the authority of the Fire and Rescue Service Incident Commander following careful consideration of specialist advice.

8B2.32 Some tunnel control rooms contain water inundation protection facilities, including tunnel portal door closers.

8B2.33 Similarly, it is not the function of the Fire and Rescue Service to operate or control these mechanisms, and their operation is the responsibility of the appropriate strategic authority or infrastructure manager.

2.9 Identify risks and hazards

8B2.34 This section should be read in conjunction with National Generic Risk Assessments for incidents in tunnels and underground.

8B2.35 At known underground locations many of the risks and hazards will have been identified in advance and relevant control measures should form part of the agreed intervention strategy as part of any pre-planning process. However Incident Commanders will need to devise and implement additional control measures where it is evident that any assumptions made as part of any intervention strategy can no longer be relied upon.

8B2.36 There is the potential for a Fire and Rescue Service to be called to a location that is unknown or unexpected in its underground size or complexity. In such situations the life risk to the public is likely to be low.

8B2.37 It would be prudent for Fire and Rescue Services to establish and maintain contact with organisations that may be able to provide assistance or advice, e.g:

- Mines Rescue
- Ministry of Defence Police
- Defence Fire Risk Management Organisation
- Home Office Police
- Health and Safety Executive
- Specialist construction engineers
- Local authority representatives.

PART B-3

Phase 3 – Planning the Response

Possible actions

Phase 3 – Actions Planning the Response	• Think • Prioritise objectives • Plan
3.1	Identify and prioritise objectives
3.2	Establishing proportionate control over the infrastructure
3.3	Formulate and transmit appropriate messages
3.4	Choose appropriate access and egress routes
3.5	Select and establish relevant cordons
3.6	Select appoint and brief appropriate safety officers
3.7	Actions of deployed crews
3.8	Determine firefighting tactics
3.9	Carry out rescues
3.10	Resolve hazardous material (Hazmat) issues
3.11	Establish effective systems for liaison

Considerations

3.1 Identify and prioritise objectives

NATIONAL INCIDENT COMMAND SYSTEM

8B3.1 Setting of objectives at tunnel or underground incidents should be approached in the same way as all other operational incidents and should observe the principles established in National Incident Command System.

COMPLEXITY

8B3.2 This type of infrastructure can be complex and present hazardous working environments, requiring a flexible approach to be adopted when planning a tactical response. The extreme conditions that can rapidly develop in this type of infrastructure, and the potential for disorientation, can make operations difficult, tiring and resource intensive. It should be remembered that even infrastructure that appears straightforward can lead to individual and crew confusion because of the repetition of features and the lack of way finder indicators.

TRAINING

8B3.3 Fire and Rescue Service Incident Commanders and Crews should use their training and knowledge, combined with local policy and procedures, to identify the hazards, risks and control measures appropriate to the incident type they are responding to. (See Generic Risk Assessment Tunnels and Underground)

INCIDENT TYPE

- **Rescue:** Rescue operations involving tunnels and underground structures can range from a single person trapped in a road vehicle, or a worker/member of the public that has become unwell or fallen into a shaft, through to extremely complex and large scale operations, involving multiple rescues and casualties, undertaken over several days.
- **Fire:** Fires can range from a small smouldering fire that presents difficulties because of the volume of smoke involved, through to protracted, resource intensive, multi agency incidents, with national or international implications on public services or military capabilities.
- **Hazardous material (Hazmat):** Incidents are likely to range from leaking valves, through to significant spillages or ruptures of goods in transit or storage. These incidents may have implications for Hazardous material (Hazmat) incident resourcing, monitoring and management

LEVEL OF CONTROL

8B3.4 The protection of the public, Fire and Rescue Service personnel, other responders in the inner cordon, and the protection of infrastructure and the environment remain the duties of the Incident Commander.

8B3.5 The first objective is to establish an environment for responders and casualties that proportionately protects against the hazards presented by the infrastructure and the incident to be confronted.

8B3.6 Incident Commanders will need to determine the level of controls to be applied to the incident balanced against the potential harm caused to the community, including those remote from the incident, but still in the affected infrastructure. This assessment may include:

- The type, severity, extent and likely development of incident
- Any immediate action required to save life
- The type of underground structure, the use it is being put to and the associated known hazards
- Passengers leaving vehicles walking onto the infrastructure and exposed to additional hazards
- Exposure to hazardous materials (Hazmats), including fumes from combustion engines, either drawn into the infrastructure by surface incidents, or released from inside
- Physical and mental distress of passengers, public or staff held within the infrastructure. Potentially made worse by overcrowding, hot conditions, failure of air conditioning or ventilation systems
- Inundation spreading to wider community, due to the topography of the infrastructure
- Limitations of Fire and Rescue Service personnel and equipment
- Widespread disruption to national infrastructure and public perception of national resilience.

3.2 Establishing proportionate control over hazards

8B3.7 When the Incident Commander has determined the incident objectives, it will be necessary to identify and establish the level of control to be implemented. The procedure for establishing control will depend on the use, hazards and associated risks for the particular infrastructure.

HAZARDS

8B3.8 Normally the specific risks for any given location will be known through preplanning and training, but in general terms the hazards likely to be encountered include:

- rapidly developing fire and high heat intensity
- difficulty in assessing the extent, nature and likely development of an incident
- difficulty moving about and applying extinguishing media owing to, for example, vehicles, available space of infrastructure, storage arrangements
- developing smoke plume and plug or other hazardous materials released and held within infrastructure
- products of the incident being dispersed into community

- running fuel fires
- contamination of water system
- limited access and egress
- potentially large numbers of people involved
- hazards, such as high voltage electricity cables, under more than one controller
- hazard from spalling of concrete
- hazard from projectile spalling where large fire has developed
- machinery or moving vehicles
- dependency on safety systems under the control of third party, e.g. tunnel ventilation, traffic movement control
- limited communication facilities
- limited fire fighting or rescue facilities
- deteriorating conditions for those trapped or engaged in rescue, owing to high heat, humidity and smoke logging
- noise from Fire and Rescue Service and other machinery, activity and equipment
- distressing scenes.

CONTROL MEASURES

8B3.9 The intervention strategy will normally establish the control of hazards by providing Fire and Rescue Service facilities and one or more of the following methods:

- Activation of local or area wide automated fire/smoke ventilation and/or suppression systems, normally by the infrastructure manager, upon discovery of an incident
- Contact with the infrastructure controller by agreed means, normally through an on-site responsible person at Silver or via Fire and Rescue Service Control, requesting remote implementation of request. (for example, over traffic movement or electrical current)
- Liaison with infrastructure's representatives and specialist Fire and Rescue Service advisors to identify community impact
- Knowledge of the infrastructures evacuation strategy and information on its progress
- Discussion with the infrastructure manager confirming the extent to which any control has been implemented. Thereby assisting the Incident Commander in identifying the safe working area within the infrastructure for operations to take place

- Use established procedures for type of infrastructure to establish appropriate safety measures (for example, road traffic or railway incident procedure) to control traffic or machinery
- Confirmation from the infrastructure managers on the status and operation of systems used to protect members of the public, staff and firefighters, for example:
  - ventilation systems
  - pressurised escape area or intervention shafts
  - vehicle control systems (for example road traffic lights set to red at either end of tunnel)
  - current status of high voltage electricity.
- Use of any of the infrastructure manager's rescue or recovery teams
- Use of survivor reception centres and staff or customer support systems.

8B3.10 Further important control measures include:

- the physical fitness of responders
- the suitability of equipment to function in a tunnel or underground environment, for example:
  - the use of appropriate rescue equipment that, for example, is not dependent on safe air, does not release carbon monoxide (combustion engines) and is not excessively noisy
  - breathing apparatus and associated equipment capable of meaningful intervention or means of transporting crews to scene
  - firefighting media and equipment capable of tackling reasonably foreseeable incidents
  - communication facilities for sub-surface operations.
- training and competency of Fire and Rescue Service personnel in identifying the specific hazards, risks, intervention strategy and operational techniques
- use of policies, tactics and equipment that mitigate the risks posed by the tunnel or underground operating environment.

3.3 Formulate and transmit appropriate messages

8B3.11 Messages to Fire and Rescue Service Control which may be subsequently relayed to the infrastructure manager require accuracy in formulation and transmission.

8B3.12 To assist with an infrastructure manager's implementation of controls over hazards to Fire and Rescue Service personnel it is necessary to include relevant details such as:

- if people are involved, or likely to become involved, in the incident

- the location of the incident
- what level of control is required over particular hazards, (for example vehicles stopped, power off)
- over what geographical area controls should apply
- nature of fire and rescue service activity being undertaken.

8B3.13 Associated details from informative messages on the nature of the incident will also provide useful information to the infrastructure manager on potential wider infrastructure and community impact.

8B3.14 It must be noted that it is the Fire and Rescue Service's expectation that no unreasonable delay should occur in the implementation of the request for appropriate and proportionate control over any infrastructure.

8B3.15 It should also be recognised by the Fire and Rescue Service that there may be wider implications to the infrastructure manager's employees or customer or community safety by the implementation of any request. Therefore messages should clearly describe the operational situation.

8B3.16 The Incident Commander and Fire and Rescue Service Control should understand that the content and accuracy of the information exchanged will inform the urgency of the request, and define the areas that the request applies to. Following a request the infrastructure manager may, without disproportionate delay, take steps to protect staff and the public who may be remote from the incident, but at risk. Examples may include people in vehicles being held, or employees of electrical companies isolating high voltage electrical supplies

8B3.17 At incidents involving infrastructure attracting an attendance from two Fire and Rescue Service areas the agreement of who is Incident Commander is essential, and would normally have been agreed by protocol. The responsibility to convey the extent and level of control implemented over particular hazards to all personnel attending is important. This should be done by locally agreed means and repeated via Fire and Rescue Service Controls.

3.4 Choose appropriate access and egress routes

8B3.18 When responding to an incident it is usually preferable to use the pre-determined intervention strategy specific for the location. The intervention strategy will consist of the method of access and egress to the infrastructure by the Fire and Rescue Service, agreed with the infrastructure managers.

8B3.19 Intervention strategies normally include locations designed or mutually agreed with the infrastructure managers for providing suitable primary and secondary Fire and Rescue Service access points, for example:

- stations
- Intervention point/evacuation point

- emergency response location
- service shafts
- stairs near portals
- tunnel portals
- ladders
- large car park areas.

8B3.20 These locations will normally provide the initial rendezvous point/strategic holding areas. Such locations may also be provided with facilities to assist access to the scene and provide facilities for firefighting and rescue operations.

8B3.21 For some incidents, particularly in older infrastructure, it may be appropriate to mobilise or request an attendance to more than one location. This will assist where:

- there is limited or no smoke ventilation or fire stopping
- there is limited or no effective communication systems.

8B3.22 There is likely to be large numbers of people involved in evacuation,

- the numbers of people involved in evacuation may significantly impact on operations
- people may not be evacuating to the instructions provided by the infrastructure manager
- information from other agencies or responders is required, and these are located elsewhere
- access should not be made until appropriate control over known hazards is implemented.

8B3.23 In all circumstances crews should receive an appropriate briefing, the details of which may include:

- tactical plan and means of implementation
- the task to be performed
- details of resources and any limitations
- the defined working area
- the path to and from the work area
- the hazards present and the level of control implemented
- evacuation signals and corresponding actions
- confirmation of understanding.

8B3.24 In all cases, when the Incident Commander is determining where and how to deploy crews and resources the type of incident and the nature of any threat, including secondary devices, will also be a contributing factor.

8B3.25 For incidents involving long distances arrangements may be put in place to use rail or road vehicles. These vehicles may consist of:

- the rail system or infrastructure manager's own vehicles
- specialist rail vehicles i.e. road/rail vehicles or trolleys.

8B3.26 Considerations that apply to attending incidents on railways should be cross referenced with other appropriate national operational guidance.

8B3.27 Any vehicle, lift or hoist that has not been designed to carry passengers should not ordinarily be used as a means of transporting crews. If a lift is provided, but it is not a designated as a fire lift, it should only be used for carrying equipment in fire situations.

3.5 Select and establish relevant cordons

8B3.28 Initiating cordons in this infrastructure will normally incorporate portals, shafts, tunnel bores, cross passages or stairs. This will normally mean that the inner cordon will be within the built infrastructure. This may be used to assist the Incident Commander in defining sectors and controlling access and egress.

8B3.29 Consideration should be given to the effect of any outlet releasing the products of the incident into the wider community. This may mean that the inner cordon could be extended to include hazard areas on the surface at points along the infrastructure.

8B3.30 It is important to communicate with other responding organisations and infrastructure managers staff the exact extent of the inner cordon to:

- protect people from tunnel or infrastructure hazards
- prevent a worsening of the incident
- preserve the scene for investigators
- define the safe working area
- assist with command and control.

8B3.31 In twin bore infrastructure it may be possible to limit the extent of the inner cordon to include only a section of the affected bore.

8B3.32 In infrastructure where smoke control facilities are provided crews may be deployed 'up wind' of the incident in order to protect them from the products of the incident, in safe air. The proximity of the crews to the incident will depend on the intensity of any radiant heat or other hazards associated with the particular use of the infrastructure.

8B3.33 For complex infrastructure, or where the infrastructure has no distinguishing way finders or where there is a lack of location indicators the use of lights, markers or barrier tape should be considered to indicate the extent of the inner cordon. Similarly these items should be considered for indicating which door or level leads to the way out along the route to the surface. This may be particularly important where the shaft is part of a larger surface building.

8B3.34 If there is any reason to believe that the way to or from the incident could cause confusion, for example because of the lack or repetition of features, then consideration of deploying suitable markers or using of guide lines should also be considered.

8B3.35 If there is the possibility of crews moving from one area directly into a hazard, for example moving traffic (this may be later in the incident when a single bore of a road or rail tunnel has been re-opened) then consideration must be given to ensuring crews cannot accidentally enter the risk.

8B3.36 Certain features of the tunnel infrastructure can be used to good effect to further define sectors. This could include:

- stations
- platforms
- tunnel portals
- tunnel cross passage doors or shafts
- marker plates
- signage
- signals
- marker posts
- vehicles/carriages.

8B3.37 Within a cordon it may be appropriate to define paths for access and egress to the scene of operations. This will assist in reducing the risk from other hazards and preserving evidence. While any Fire and Rescue Service cordon is in place care should be taken to ensure access and egress is controlled and the number of personnel operating in the area is kept to a minimum.

8B3.38 As soon as is reasonably practicable a reduction in cordon size should be considered, in consultation with other agencies. Consideration should also be given to any reasonable requests for adjustment to Fire and Rescue Service operations in order to provide restoration of service when it is safe to do so.

8B3.39 The cordon at an incident in this infrastructure is likely to be linear in nature, and may cover a considerable distance. It may be necessary within or alongside the cordon to position staging posts, marshalling areas or command facilities.

8B3.40 Where smoke control ventilation is provided, it is normally the case that doors should be kept closed as far as possible. This will assist in ensuring the pressurisation is maintained at the optimal level. The use of any door wedges should be considered bad practice.

8B3.41 Evacuation from within any cordon can initially involve large numbers of people. The responsibility for their evacuation rests with the owner or infrastructure manager.

8B3.42 Incident Commanders should seek information from the infrastructure manager to measure how any evacuation is progressing. This will include identifying items such as, whether a roll call has taken place, or the existence of safe areas/refuge rooms.

8B3.43 Special consideration should be given to identifying and prioritising people who are, for whatever reason, at greater risk such as impaired mobility.

8B3.44 If there is any concern relating to the management of the evacuation the Incident Commander should consider providing assistance or taking command of the evacuation, with the infrastructure manager's support.

3.6 Select, appoint and brief appropriate safety officers

8B3.45 Any safety officers appointed should be briefed on the specific nature of the hazards they are responsible for monitoring, and the actions to be taken if additional control measures are required. Safety officers should also ensure that the position they take to monitor operations does not place them at risk from the range of hazards listed above (section 3.2).

8B3.46 Part of the assessment of whether to deploy safety officers and where they should be positioned to perform their duty should be subject to the following considerations:

- speed and stopping distances of any vehicles
- distance from the scene
- complexity of the location
- ventilation and lighting conditions
- communication and evacuation methods to be used
- the audibility of any message or signal
- the noise level at the scene and the size of operation
- the risk to the safety officer
- the availability of Infrastructure staff to undertake safety role
- the ability and speed of infrastructure managers to remotely remove the hazard.

3.7 Actions of deployed crews

8B3.47 Fire and Rescue Service personnel must be aware that in addition to any control measure implemented to protect the scene, a general awareness of the hazards likely to be involved must be maintained and appropriate control measures implemented.

8B3.48 At the scene of an incident those undertaking operations at a bronze level will be responsible for reassuring the public and taking measures to protect people from further harm. This may be achieved by:

- providing reassurance, information and instruction to members of the public
- relieving members of the public or infrastructure staff from any spontaneous rescue operations, when reasonable to do so
- keeping alert. 'Stop, look, listen' before moving about the infrastructure
- be aware of safety signage or information provided for Fire and Rescue Service crews
- providing an initial assessment of the situation and regular updates to the Incident Commander
- informing the Incident Commander of any control measures implemented by infrastructure staff or vehicle/machinery operators
- use equipment and facilities designed for access and fire and rescue service purposes
- making an assessment of surrounding hazards before using Fire and Rescue Service resources
- after power off avoid unnecessary contact with electrical equipment
- coordinate with specialist rescue personnel employed by infrastructure managers
- remain aware of the situation and any changes you make to the scene, to assist with any future investigation
- following significant incidents, it may be useful for individuals to record observations and any actions taken as soon as is reasonably possible.

3.8 Determine firefighting tactics

8B3.49 When undertaking and directing firefighting operations consideration should be given to:

- difficult working conditions that may rapidly become untenable if application of extinguishing media is delayed
- use of relevant firefighting procedures agreed in the intervention strategy, appropriate to the hazards and risks present
- proximity of crews, firefighting jets and equipment to hazards

- the resource intensive nature of these incidents
- long travel distances to bridgeheads and marshalling areas
- the welfare issues around resourcing such incidents including:
  - food and fluid replacement
  - toilet facilities
  - reliefs and shift structures
  - use of relief crews from areas not familiar with infrastructure.
- implementing measures to expediently assess the damage caused to the infrastructure, for example arches, tunnels, key elements of infrastructure
- hose lines traversing rail or roadways, not protected by control measures
- maintaining an awareness of the defined safe working area, sector and cordons
- use of any tactical plans or built in facilities
- ascertaining the status of any automatic system to assist firefighting
- establishing and maintaining agreed communication procedures.

8B3.50 The incident commander will need to identify:

- the extent of the infrastructure
- the location of the incident in the infrastructure
- what is involved for example train, coach, machinery, and the location of the fire/rescue therein
- the size and likely development of the fire and/or rescue operation
- where are the staff and members of the public
- the location of others who may become involved, for example in other vehicles or on platforms
- the direction of ventilation
- any gradient in tunnels
- alternative access and egress options.

8B3.51 Any work is likely to be demanding and additional resource and welfare requirements are likely to occur.

8B3.52 Some examples of firefighting tactics that may be used in various types of infrastructure are shown below:

Picture 1: Standard procedure for an incident occurring in a single bore tunnel without access tunnels

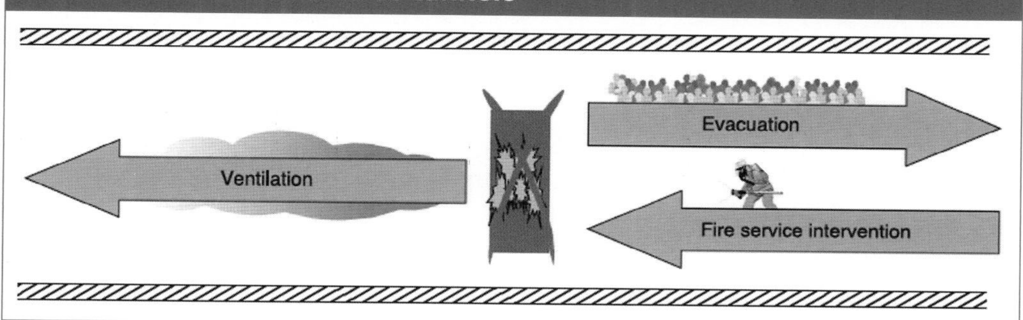

Picture 2: Standard procedure for dealing with an incident occurring in a single bore tunnel via an access tunnel

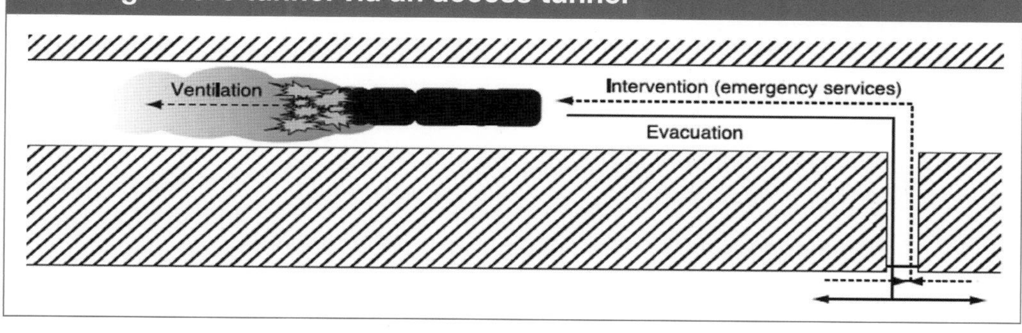

Picture 3: Standard procedure for dealing with an incident in a twin bore tunnel via the unaffected bore and crossovers

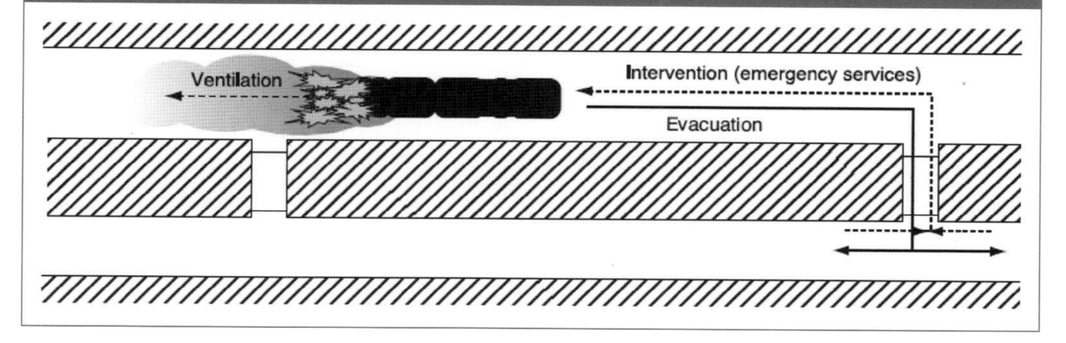

Picture 4: Standard procedure for dealing with incidents in a twin bore tunnel via a service tunnel

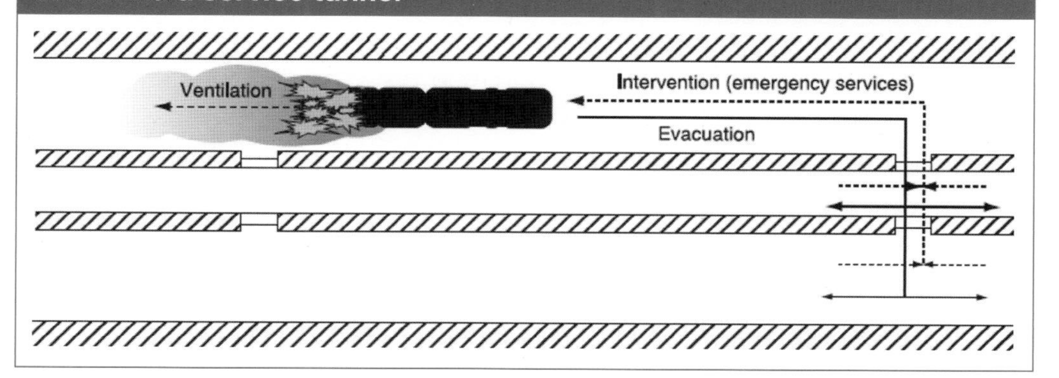

Diagrams from *Handbook of Tunnel Fire Safety* published by Thomas Telford Publications

Common principles

8B3.53 The above are very general descriptions of firefighting tactics. The necessity to keep any intervention or evacuation strategies simple in order to reduce the chance for confusion or error in the implementation of firefighting tactics should not be underestimated.

8B3.54 Fundamentally the Incident Commander should be looking to:

- remove smoke and fumes from the hazard area as quickly as possible
- ensure the evacuation of the public to a place of safety has been implemented
- implement rescue and firefighting operations from bridgeheads, removing victims to a place of safety and care.

3.9 Carry out rescues

8B3.55 Methods for rescuing should reflect, as far as possible, the standard practice for the type of infrastructure. For example a road incident should employ standard road incident procedures.

8B3.56 Incident Commanders and crews should use appropriate tools and equipment, with particular attention to combustion engine powered equipment or vehicles. As these may be unable to function, due to oxygen deficient atmosphere, or add to noxious gasses and noise inside the infrastructure, to the detriment of those needing aid or rendering assistance.

8B3.57 The positioning of air inlets outside the infrastructure should be identified to ensure that exhaust gasses are not entering the ventilation system.

8B3.58 Noise can also cause unnecessary harm and interfere with communication or any emergency signal.

8B3.59 At large scale or protracted incidents the use of regional or national resources may be appropriate. Such resources include:

- Urban Search and Rescue
- Incident Response Unit
- Specialist equipment.

8B3.60 For incidents involving rescue and recovery, consideration of hygiene and infection control should be considered. This is particularly the case in hot and humid environments involving live rescue efforts where deceased casualties may be co-located, or where there is a risk of zoonoses.

8B3.61 In such conditions crews may wish to remove items of personal protective equipment for comfort and ease of access. Incident Commanders or Safety Officers should liaise with health professionals to assess the risks and appropriate control measures. These may include:

- provision of temporary ventilation
- post incident health monitoring
- vaccination
- dressing any cuts or grazes before entering cordon
- preventing access to those with cuts or grazes.

8B3.62 When moving through sectors or areas with differing hazards it will be necessary for crews to be rigged to the highest suitable level of protection.

8B3.63 When searching or rescue operations are underway at larger incidents, care should be taken when assigning identifiers to sectors or carriages involved. It is important to ensure that any numbering system applies to the correct area from either side of the incident.

8B3.64 Any identification system used should be shared with other agencies present.

8B3.65 The evacuation of passengers, including those who may have restricted mobility, that do not require rescue, is the duty of the infrastructure managers.

8B3.66 Incident Commanders should be aware of the evacuation strategy for the infrastructure or location, and should enquire how any evacuation is progressing.

8B3.67 Infrastructure under construction or classified as mines, sewers, confined spaces, and some railways may have specialist teams, either on site or available. These may be either responsible for rescue operations in their own right or to assist Fire and Rescue Service. Incident Commanders should seek an early appreciation of the availability of any such resources, whether they are deployed, their responsibilities and capabilities.

8B3.68 The Incident Commander will need to identify:

- the location of those requiring rescue within the infrastructure
- the size of the fire/hazardous material release/rescue operation
- actions required to save the lives of the victims
- the risk to those who enter the infrastructure
- capability of equipment
- further resource requirements
- best method of clearing smoke or fumes from infrastructure.

8B3.69 When developing tactics for rescues Incident Commanders should make full use of any on site or on call specialists, or their rescue personnel and equipment, to provide 'expert' information and opinion that is relevant to the specific infrastructure.

3.11 Resolve hazardous materials (Hazmat) issues

8B3.70 Standard hazmat procedures will be broadly applicable to any hazmat incidents, further information on this subject is available via the relevant national generic risk assessments and the national operational guidance for incidents involving hazardous materials.

8B3.71 Hazmat incidents occurring will predominantly be as a result of:

- the nature of goods being transported or materials stored
- release of natural gasses, normally during construction or mining
- hazardous materials being deliberately introduced.

8B3.72 The nature of the incident will inform the Incident Commander's actions when determining response in terms of resources, deployment and crew safety.

8B3.73 Incident Commanders should ordinarily consider the effect that the incident may have on the occupants, wider community and environment, examples include:

- the effects of any ventilation system exhaust carrying contaminates into the community
- the potential for secondary contamination of locations such as ambulances and hospitals
- the effects of incident runoff on near-by water courses, particularly in more remote/rural locations
- the gradient of the infrastructure allowing contaminants to flow away from the scene
- the type of interceptors or drainage installed, if any
- the limitations of any installed detection identification and monitoring equipment.

3.11 Establish effective liaison systems

8B3.74 The management and resolution of incidents in tunnels and underground structures often requires the combined efforts of a number of agencies. It is therefore essential that effective liaison is established at an early stage to ensure that the priorities and hazards relating to the different agencies and responders are effectively managed.

8B3.75 Silver meetings are particularly important. The early and regular exchange of information on the progress of emergency responders, assessment of the impact of an incident on the community and industries involved along with comprehensive reviews of safety issues can help to coordinate response and reduce the impact of the incident. In addition to normal protocols for silver meetings, the following points should be considered at incidents:

- the likely affects or implications of the incident should be discussed and the effects reduced or mitigated

- identifying if any responders have designated the incident as a 'major incident'
- identifying any communications issues, assessing the impact and any mitigation that can be applied
- confirmation of the extent of the cordon in place and the current status of specific hazards
- appropriate reduction of the cordon, when safe to do so, to reduce the community or industry impact of the incident
- the progress of any industry managed evacuation and the impact on Fire and Rescue Service operations
- the ownership of the infrastructure and its uses should be identified and communicated to all agencies. For example a road tunnel may be the responsibility of one of the following:
  - strategic transport authority
  - local authority
  - private company
  - a Crown organisation
  - a rail infrastructure manager.

PART B–4

Phase 4 – Implementing the response

Possible actions

Phase 4 – Actions Implementing the response	• Communicate • Control
4.1	Implement effective control measures
4.2	Implement effective firefighting and rescue operations
4.3	Communicate the tactical plan
4.4	Implement deliberate reconnaissance to gather further information
4.5	Communicate with other agencies
4.6	Controlling the tactical plan

Considerations

4.1 Implement effective control measures

8B4.1 Incident Commanders should ensure that appropriate and proportionate control measures are requested from the relevant infrastructure managers, using the most expedient method. Normally this will be requested and confirmed by radio transmission, via Fire and Rescue Service Control.

8B4.2 It would normally be desirable to obtain confirmation from the infrastructure managers that the control measures have been implemented before committing personnel into the hazard area.

8B4.3 In extreme circumstances personnel may be committed before confirmation has been received. In such circumstances additional, temporary, control measures must be applied. For example in areas with extensive, old, complex or high risk infrastructure a suitable sub surface procedure would provide additional control measures for crews.

8B4.4 Recording those committed into the infrastructure or into the inner cordon is essential.

8B4.5 Contact with any other Fire and Rescue Service resources attending alternative access points should be established and maintained to inform the Incident Commander's risk assessment on hazards, progress and risk benefit analysis.

8B4.6 In addition to Safety Officers being appointed for conventional Fire and Rescue Service purposes at these incidents a Safety Officer may be appointed to give warning of hazards specific to the infrastructure or its use.

8B4.7 There will be certain types of infrastructure, for example high speed train tunnels or high voltage electricity cable tunnels, where it is unlikely that temporary control measures will provide sufficient protection from the hazards present.

8B4.8 Fire and Rescue Service crews should receive an appropriate briefing on the actions to be taken and the firefighting, rescue or environmental protection tactics to be employed. The briefing may include:

- Information on the significant hazards and risks likely to be present, and control measures
- Warning indicators or automatic evacuation signals that may be present, and the crews subsequent actions
- The areas where operations are to be conducted, where appropriate control measures have been applied
- The areas that are not to be entered or for any work to be undertaken
- Information on fixed installations provided, for example the opening of any valves along pipe work, required to supply firefighting media, or how they provide protection
- The role of other agencies working with Fire and Rescue Service personnel
- Instructions on moving fatalities. This will normally involve leaving those beyond assistance in situ, unless they are likely to impair wider rescue operations. If there is a strong probability that important evidence may be lost if the fatality is left in place.

8B4.9 When working alongside any other agency the necessity to communicate identified hazards and the control measures implemented is essential. It is also important to communicate those hazards that relate to Fire and Rescue Service operations and environment, which other agencies may not be familiar with, and that may present greater risk at these locations.

8B4.10 As mentioned before, working in such an environment can cause considerable strain. The establishment of suitable first aid facilities, alongside any entry control, tally or entry recording point, to support committed crews, would be good practice.

4.2 Implement effective firefighting and rescue operations

8B4.11 Hazards usually faced by responders are made more challenging and present significantly greater risks by the characteristics and use of the tunnel or underground structure.

8B4.12 The type and use of the location and the risk to human life, property and the environment will assist the Incident Commander in determining the extent of the duty placed on the Service and the extent of the service the Fire and Rescue Service will provide.

8B4.13 Many Fire and Rescue Service personnel will not have experience of the arduous nature of such operations, particularly those associated with fire fighting operations in tunnels and underground locations. These factors need to be considered by the Incident Commander when implementing the tactical plan. Further considerations may include:

- Deployment of appropriate resources and equipment, normally within an intervention strategy. This may include specialist vehicles or equipment, specialised procedures such as sub-surface, or the use of national or regional resources
- Utilisation of facilities to assist operational tactics, including:
  - use of cross passages, shafts and lobbies to form bridgeheads, or gain easier access to an incident
  - use of automatic smoke control to approach incident from 'clean air'
  - ensuring effective communication with fire control and detection systems, panels and control rooms.
- Identify alternative locations for access and egress
- Ensuring the maintenance of crews' means of escape
- Ensuring there are adequate support resources available to initiate and maintain a sustained and uninterrupted operation
- An appropriate briefing to personnel entering the inner cordon
- Any tactical plan needs to be realistic and achievable, balancing risk against benefit
- Operations may involve joint working with other agencies including HART or specialist police departments and the need to understand the function and parameters of their operations and any associated risks
- Utilise and coordinate the specialist advice, staff and equipment provided by the infrastructure manager to support Fire and Rescue Service operations.

4.3 Communicate tactical plan

8B4.14 The plan will need to be communicated to all Fire and Rescue Service personnel on site. This will include any attendance sent to an alternative access point, elsewhere on the infrastructure.

8B4.15 Discussion with other responders working within the inner cordon should also take place. This discussion should consider coordination of key issues such as:

- common evacuation signals
- inter-service communications that are available
- any potential impacts on each others operations.

8B4.16 Any tactical plan should be structured and communicated within the standards set by the incident command systems and local major incident procedure protocols.

8B4.17 If reasonably practicable the infrastructure manager should be advised of the intended tactical plan. It must be made clear that there should be no changes to any automatic ventilation or fire control systems settings, for the duration of the incident or, until the Incident Commander requests such a change.

8B4.18 Incidents led by other agencies, for example police, will require careful inter-agency liaison to ensure the nature of the incident and the associated risks of any parallel operations are understood and appropriate control measures are implemented.

4.4 Implement deliberate reconnaissance to gather further incident information

8B4.19 Deliberate reconnaissance at incidents on this type of infrastructure is particularly important. There may be very few indicators of the type and severity of incident developing.

8B4.20 Care will be required when entering and moving around the infrastructure. Quite serious incidents can be developing without any obvious indication, and may only become apparent when a door is opened. By which time the crew may be exposed to the full hazard. This is in addition to the normal operating hazards associated with the use of the place

8B4.21 Full use of any intervention facilities should be taken, and care in the selection of personnel is appropriate. Selection should involve those who have current knowledge, skills, training and fitness levels appropriate to the hazards likely to be encountered at a reasonably foreseeable incident.

8B4.22 In infrastructure providing high standards of protection from the effects of an incident the use of on-site staff may be considered to speedily direct crews to a bridgehead or lobby close to the incident. Alternatively the use of any Fire and Rescue Service or infrastructure manager's vehicles should be carefully considered to allow personnel and equipment to the scene.

8B4.23 Crews involved in reconnaissance should ensure that they have appropriate protective equipment and communications facilities to enable their safe withdrawal from an incident.

8B4.24 The crew will also ensure that they maintain the integrity of any protected route they are using to ensure their egress will not be impaired. The potential to become confused or disorientated in even simple infrastructure should not be underestimated.

8B4.25 On discovery of information the crew must provide the Incident Commander with quality information on the type and severity of any incident, before becoming involved with any other operational activity.

8B4.26 Reconnaissance information may also involve the use of information provided by:

- control rooms, either on site or central to the wider infrastructure
- CCTV
- other emergency services
- staff communication systems
- on site detection equipment (for fire, smoke, hazmats- including natural gasses)
- plans boxes
- pre-defined tactical plans
- infrastructure managers, staff or responsible person at silver
- warning signs and signals
- witnesses.

8B4.27 Assistance in interpreting the information obtained by reconnaissance my be obtained from:

- specialist Fire and Rescue Service Officers including hazardous material (Hazmat), Inter-Agency Liaison Officer (ILO)
- infrastructure managers, staff or responsible person at silver
- specialist engineers or technicians
- de-briefing crews and information.

4.5 Communications with other agencies

8B4.28 Incidents involving tunnels and underground structures can present some significant challenges for communications infrastructure. There are no simple rules that can be applied and the communications problems of individual locations need to be considered from both the dynamically deployed operational Fire and Rescue Service and multi-agency perspective and from the fixed systems perspective as supplied in many modern structures.

8B4.29 Incidents often require a multi agency response to achieve a satisfactory conclusion. Historically communications both internal and external have been identified as areas of weakness in post incident investigations and debriefs. Therefore Fire and Rescue Service Incident Commanders should consider carefully their methodology for communication with other agencies (Category 1/2 Responders). Areas for consideration will be:

- AIRWAVE radio system utilising interagency radio channels

- danger of reliance on mobile telephone networks
- the use of field telephones between emergency service control vehicles
- the use of runners if appropriate
- the use of inter agency liaison officers
- the use of any mutually agreed method to overcome local difficulty
- the use of Silver meetings to confirm incident situation and communication inter service communications structures and limitations.

8B4.30 Information is one of the most important assets of incident management; information must be gathered, orders issued and situation reports received. The needs for communicating with other agencies must be assessed and provided for.

8B4.31 The Incident Commander will need to:

- establish communication with the tunnel operator.
- establish communication links with Brigade Control
- establish command of the incident in accordance with the Incident command system allocate UHF radios, assign channels and agree on call signs
- establish communications with other agencies
- establish and monitor communications within the tunnel environment.

4.6 Controlling the tactical plan

8B4.32 Once the tactical plan is in place, the Incident Commander must ensure that effective arrangements are in place to monitor the implementation, application and progress of the plan to ensure the objectives are being met. This will include the establishment and maintenance of:

- an appropriate command structure
- effective safety management systems
- appropriate communications systems
- effective arrangements for liaison.

PART B–5

Phase 5 – Evaluating the response

Possible actions

Phase 5 – Actions Evaluating the Response	• Evaluate the outcome
5.1	Obtain and utilise specialist advice
5.2	Assessment of safe systems of work
5.3	Evaluate the effectiveness of the tactical plan
5.4	Consider the appropriate and timely reduction of the size and impact of cordons

Considerations

5.1 Obtain and utilise specialist advice

8B5.1 At known infrastructure there will normally be arrangements in place to obtain specialist advice on the structure, its contents and processes.

8B5.2 For premises that are used as roadways or tunnels the procedures for obtaining advice from the infrastructure manager will normally be the same for locations not in tunnels or underground.

8B5.3 For premises that are unknown the local authorities will be able to provide information on the stability of the structure.

8B5.4 The Incident Commander will look to identify relevant authorities and request attendance of specialist advice as required.

5.2 Assessment of safe systems of work

8B5.5 Safe systems of work that are implemented as control measures to protect Fire and Rescue Service personnel and possibly other responding agencies working in the same location, should be continually re-assessed with consideration of the following:

- the information received from those listed in 4.4 confirms the justification of continued Fire and Rescue Service activity
- the likely loss or impact caused by an incident
- the potential for escalation of the incident, likely involvement or spread of fire, hazardous materials

- the potential for a delay between changes being made outside the infrastructure and those changes being realised underground
- a change in weather or any flood water conditions may have an affect on the suitability of the safe systems of work
- the stability of the underground structure as the incident develops, being affected by Fire and Rescue Service operations or the result of fire/crash damage
- the functionality of any fixed installations and support equipment, including Fire and Rescue Service communication facilities.

5.3 Evaluate the effectiveness of the tactical plan

8B5.6 With all tactical plans there will need to be a continuous review of the priorities and objectives of the plan balanced against the risks being taken by Fire and Rescue Service personnel. Undertaking this review the following questions may be considered:

- Is crew safety and welfare being maintained?
- Are the risks being taken by fire and rescue service personnel proportional to the benefit?
- Have comprehensive analytical risk assessments been completed and appropriate control measures implemented?
- Are the resources appropriate and adequate to achieve the tactical plan?
- Has there been a change in the level of risks pertaining to the incident for example, rescues being carried out?
- Has any review of the plan required a change to the level or extent of control measures?
- Have the operational tasks achieved the tactical plan, if not why not and what needs to be altered to achieve the tactical plan?

5.4 Consider the appropriate and timely reduction of the size and impact of cordons

8B5.7 Obtain information on the progress of Fire and Rescue Service actions and any evacuation of members of the public by the infrastructure managers

8B5.8 Establish early silver meetings to:

- develop effective joint plans to mitigate the impact of the incident and
- agree handing over process to the appropriate organisation, agency. or company
- develop any media strategy and provide information to the public

8B5.9 Gathering information from responders and specialist advisors can assist an Incident Commander to evaluate the response this may include:

- Safety Officers
- Hazardous Material and Environmental Protection Officers
- Inter agency liaison officer
- Press liaison officer
- Infrastructure manager
- Infrastructure engineers
- Other emergency service responders on-site
- Health and safety professional or other statutory investigators.

PART B–6

Phase 6 – Closing the incident

Possible actions

Phase 6 – Actions Closing the incident	
6.1	Scaling down Fire Service operations
6.2	Handover/ownership of incident
6.3	Facilitate debriefs
6.4	Facilitate post incident reporting
6.5	Maximise learning

Considerations

6.1 Scaling down Fire Service operations

8B6.1 This is an important phase of the incident and statistically a phase when accidents and injuries are prevalent. There is a need to maintain effective command and control throughout this phase of the operations, which is likely to include:

- continued dynamic management of risk and a record of incident command decisions
- scene preservation in conjunction with advice from police/Her Majesty's Inspectorate of Mines/Health and Safety Executive/Rail Accident Investigation Branch/Office of Rail Regulation
- decontamination of equipment and personnel
- personnel welfare
- safe recovery of Fire and Rescue Service equipment.

6.2 Ownership/handover of incident

8B6.2 As an incident moves from the operational phase to the closing phase there may be a need to handover the incident to another agency.

8B6.3 It will be necessary to identify on-going ownership of the incident scene, i.e. Highway Agency, local authority, private enterprise, railway authority, in order that correct handover procedures can be put in place.

8B6.4 Where a handover takes place from the Fire and Rescue Service Incident Commander to a responsible person from another agency such as, police, mine owner, infrastructure manager or the Rail Accident Inspection Branch there must be a full and thorough exchange of information which should be recorded at a silver meeting as part of the ongoing risk assessment process and should include the following:

- the current Incident Commander
- the identification of the responsible person taking over the incident
- the risk assessments in place
- safe systems of work being employed
- actions that have been taken including rescues and number of casualties, firefighting etc.
- what actions are currently taking place
- any personnel still deployed, and what agencies they are from
- any equipment still deployed
- location of any hazardous materials
- hazardous or unsafe structures
- environmental considerations
- contact details of relevant agencies that may be required to bring the incident to a satisfactory conclusion.

6.3 Facilitate debriefs

8B6.5 As with similar types of major incident, the Incident Commander will need to ensure the relevant records and information are made available for internal, inter service and inter agency post incident debriefs, which may include:

- on scene hot debriefs
- structured Fire and Rescue Service internal debriefs
- structured multi agency debriefs
- critical incident debriefs (trauma aftercare).

6.4 Facilitate post incident reporting

8B6.6 Most incidents will be subject to some degree of post incident reporting. The extent and detail of any reporting will depend on the scale and severity of any incident. Compilation and circulation of multi-agency major incident reports may be determined by the Strategic Co-ordination Group in line with National Policing Improvement Agency guidance for emergency procedures.

8B6.7 Internal reports, logs or other documentation may be disclosable and may be used in coroners or criminal court proceedings. Incident Commanders and fire and rescue services should consider the need for the following to be created and maintained during any incident and make appropriate arrangements for the security and availability of this information following any incident. Information sources may include:

- contemporaneous notes and/or statements from Fire and Rescue ervice personnel
- continuous record of Fire and Rescue Service mobilisations and messages
- decision logs
- internal Fire and Rescue Service investigations and reports.

6.5 Maximise learning

8B6.8 Fortunately serious incidents involving Tunnels and Underground structures are rare and therefore when these occur fire and rescue services should seek to maximise the benefits to the Fire and Rescue Service as a whole. Fire and rescue services should consider the following as opportunities to measure and benchmark performance, identify potential for improvements and share lessons learned:

- national operational guidance
- Fire and Rescue Service intervention strategies
- Fire and Rescue Service policies and standard operating procedures
- Fire and Rescue Service training
- equipment failures and successes
- lessons learnt and shared with other authorities and Fire and Rescue Service.

Part C
Technical considerations

PART C–1
Introduction

8C1.1 An overriding principle for fire and rescue services to consider when planning for and responding to incidents in tunnels or underground structures is that these locations may be classified as "confined spaces" under the Confined Spaces Regulations 1997. These regulations require that fire and rescue services carry out suitable and sufficient assessments of the risks associated with all work activities so as to determine necessary control measures. In line with the Health and Safety Executive high level statement (March 2010) control measures and safety systems should take account of the wider context of Fire and Rescue Service operations and facilitate the delivery of an effective and realistic service.

8C1.2 The management of an operational incident in a tunnel or underground environment will have a number of features that Fire and Rescue Service personnel should be aware of in order to meet their duties efficiently, providing an effective service to the community.

8C1.3 Unlike many other types of incident the response will not be solely governed by the quantity of any material burning, any hazardous materials involved, or the number of people requiring assistance – for whatever reason. It is quite possible for an incident that would be relatively minor in a surface environment to pose significant challenges to Fire and Rescue Service Incident Commanders and crews.

8C1.4 Smoke or fumes are likely to be confined to the infrastructure, until released by either natural or mechanical ventilation methods. Heat, smoke or fumes may travel a significant distance from the source of the fire or release. This may present challenges in terms of, for example

- breathing apparatus deployment and logistics.
- hose lines
- transporting equipment and personnel
- continuity and quality of communication
- ensuring security of safe access and egress routes for firefighting crews.

8C1.5 For incidents involving fire or gases the fire size or quantity of hazardous materials involved is not necessarily an indicator of the level of risk presented by the incident.

8C1.6 Hot gases may not be able to escape from the infrastructure. Radiant heat will be intensified as the infrastructure returns the heat, and intensifies the thermal dynamic, back onto the incident. These factors will reduce the tenability of the area and increase the rate of burning and fire spread.

8C1.7 The fabric of the structure may be affected, particularly by direct contact with flame or prolonged exposure to heat. If the fabric of the structure has not been protected from spalling or thermal stress the likelihood and extent of damage occurring is increased.

8C1.8 The application of extinguishing media requires careful consideration in order to prevent spalling occurring, and crews should consider the affect that hot gasses and steam may have on any person that may be affected, particularly above the incident, at a ventilation point or even some distance along a passage or large compartment.

8C1.9 Increasingly modern construction has incorporated methods for the control and removal of smoke, and mitigating the effects of the considerable heat generated. Such controls may release contaminants and smoke some distance from the seat of the incident, to open air, potentially affecting wider communities.

8C1.10 Tunnels and underground locations may have some method of drainage or gradient, which will present the potential for wider environmental contamination or flowing fuel fires.

8C1.11 The location where partner agencies establish their operations may be further from the scene then normally the case. This may have implications for Fire and Rescue Service resources, partner agency capabilities, and impact on inter agency liaison and communications.

8C1.12 The environment can present further hazards such as noise and fumes from equipment. The noise may hinder operational communications, and fumes could cause harm to responders or adversely affect those in need of assistance.

8C1.13 The survivability of any victims will depend on the familiarisation of crews with the infrastructure and the ability to quickly implement appropriate controls and commence an intervention. Delay will have a detrimental effect and could allow an untenable environment to develop.

8C1.14 Some locations may have special measures in place for the rescue of occupants. Such locations include mines and premises using pressurised working. It is essential that Incident Commanders are fully aware of the statutory responsibilities placed upon the infrastructure managers at such locations. This will ensure that the fire and rescue services provides an appropriate response, if called to assist.

PART C–2

Tunnels

General

8C2.1 Tunnels are required where the route of the transport or utility system must pass under a geographical feature (river or mountain etc), a man-made feature (other tunnels or underground structures) or to keep the transport or utility system hidden for cosmetic, environmental or social reasons.

8C2.2 For the fire service, they pose particular hazards due to the:

- construction features
- wide variation in the provision of Fire and Rescue Service facilities
- inherent fire loading of the structure and facilities
- fire loading of the transport or utility system using it
- access to large numbers of the public (who may be unfamiliar with the location)
- effect on ventilation of the access structures
- communication problems.

8C2.3 Fire and rescue services should note that the ends of tunnels can be in different fire and rescue service areas, different UK administrations (the Severn Tunnels between England and Wales) or even different countries (the Channel Tunnel between the UK and France). This requires effective liaison and pre-planning to ensure different fire and rescue services are conversant with the other's procedures and capabilities and there is a joint operational procedure that both can support.

Tunnel construction methods

8C2.4 There are a number of methods used for the construction of tunnels. The construction method used to construct a tunnel must be appropriate to the ground conditions if the tunnel is to be stable and waterproof.

8C2.5 The risks from tunnels will vary depending on their age and quality of construction and maintenance, such as molten droplets of jointing materials or spalling. The spalling of the concrete is caused when the internal pore pressure is raised. When the concrete is subjected to fire the moisture in the concrete turns to a vapour and increases the pore pressure. The resulting internal pressure causes the concrete to spall. This reduces the structural strength of the concrete and produces an additional hazard to those making an intervention.

8C2.6 In fires where significant heat has developed there is an additional risk from projectile spalling of large pieces of concrete. This is particularly hazardous when it occurs above head height. In newer construction polypropylene fibres may be added to the concrete mix. In a fire these fibres will shrink, providing a cavity for water gasses to escape to, considerably reducing spalling.

Cut and cover tunnels

8C2.7 These tunnels, in the simplest of terms, are constructed by digging a trench and placing a slab over the top. Cut and Cover tunnels are constructed just beneath the surface and are all, at some stage during construction, open to the surface.

8C2.8 They consist of some form of support to the ground that forms the walls of the tunnel. This may be sheet piles, sometimes with an inner skin of in situ concrete, but could be bored concrete filled piles (usually interlinked). Alternatively, in a similar manner diaphragm wall may be used. Diaphragm walls are constructed by digging a trench, placing a reinforcement cage and then filling it with concrete. This then sets and becomes the wall of the tunnel.

8C2.9 Cut and Cover tunnels may be constructed top-down when the roof slab is constructed first, to strut the walls at the top and then the tunnel is excavated beneath and within the walls. Alternatively they can be constructed bottom-up, where the walls are strutted during excavation to retain them until a bottom slab and a roof slab have been constructed. The tunnel's roof structure will provide structural support to the walls and therefore the integrity of its configuration is essential. The roof can be either pre-cast or in-situ concrete planks and plays a vital role in the stability of the tunnel.

8C2.10 The risks presented to operational personnel from these types of structures include:

- The potential weakness of the tunnel roof slab to support the weight of a fire appliance or significant Fire and Rescue Service response. This will be a greater risk during construction when structural supports may not be set or traffic management controls may not be in place

- There may be limited 'tallying' of workers committed to the construction area as 'mining' controls may not be considered appropriate
- Explosive spalling of roof beams when exposed to excessive heat. (This is normally mitigated by fire protection measures).

8C2.11 Cut and Cover tunnels are mainly used for transport purposes; they may be single or twin bore, they may be linked to a wider tunnel network and are often used as an approach to bored tunnels. Typical examples are the M25 Holmesdale tunnel or the Limehouse Link Tunnel.

Bored tunnels

8C2.12 Bored tunnels, as the name implies, are constructed entirely surrounded by ground. In fact, they rely on being surrounded by ground for their stability. The method of construction of the tunnel is dictated by the ground conditions, whether the ground is stable as excavated, whether it is water bearing, and whether the excavated ground will remain stable.

8C2.13 These tunnels are 'driven' through the soil, under the surface, either by mechanical, manual or explosive means.

8C2.14 Hazards associated with these types of tunnels include:

- very long travel distances
- long 'dead end' conditions, especially during the construction stage.

8C2.15 Examples of bored tunnels would include all the deep tunnels of the London Underground (the subsurface lines are mainly cut and cover), the Channel Tunnel and the Channel Tunnel Rail Link. Utility tunnels such as the Cable tunnels beneath London and the London Water Ring Main provide further examples of bored tunnels.

Immersed tube tunnels

8C2.16 In instances where a road or railway crosses a river or body of water, an immersed tube tunnel may be used as an alternative to a bored tunnel.

8C2.17 Essentially, this is a concrete tube, usually of rectangular section, which is laid in a shallow trench on the sea or riverbed and then covered over.

8C2.18 The approach gradients to this form of tunnel are much lower than for bored tunnels, this provides a great advantage. It is also easier to provide a bigger space and there is little interface with the geological conditions beneath the sea or riverbed. It does however require suitable bed conditions and suitable sea conditions.

8C2.19 The construction method will vary depending on length and the waterway being crossed. In the construction stage the main emphasis will be marine in the operational stage it is like any other tunnel but has a depth of water above it.

8C2.20 Immersed tube tunnels are relatively rare in the UK but the Conway Tunnel the new Tyne Tunnel and the Medway Tunnel are good examples.

8C2.21 Hazards associated with incidents in this type of infrastructure include failure of gasket seals. This is more significant when situated in areas with high water table, rivers or lakes.

Road tunnels

8C2.22 Tunnels used for road transport range from a short, covered length of road to large-scale tunnels, extending over 1 mile in length. Tunnels over 500 metres in length on the TERN (Trans European Road Network) are legislated by the Road Tunnel Safety Regulations 2007 (amended in 2009). This legislation obliges the tunnel operator to fulfil certain duties and have systems of work in place to ensure the safe operation of the tunnel. This includes having a designated Tunnel Manager, Safety Officers and to carry out risk reduction measures. Other road tunnels are generally designed to meet the specifications of the Highways Agency Manual for Roads and Bridges.

8C2.23 Road tunnels will be either uni-directional (traffic only travels in one direction in the bore) or bi-directional (where traffic travels in both directions within the same bore). Where two bores are present and each is usually uni-directional, this allows one bore to be closed (for an emergency or routine maintenance) and the open bore to be operated bi-directionally. Where a tunnel has two bores, there will normally be cross-passages between the bores with smoke resistant doors at either end to enable evacuation of the public to a safer environment and also allow firefighting crews to use the unaffected bore to gain access to a fire via the cross-passage doors. Some have a length of pipe with male and female instantaneous couplings at each end to supply water to the affected bore, without wedging the smoke resisting doors open.

8C2.24 One constant feature of road tunnels is the lack of supervision for road users. This requires effective arrangements and signage to be present to enable evacuation from the tunnel if needed.

8C2.25 Construction features and risk reduction measures vary greatly, according to the risk posed by each structure. Therefore an overview is given of the features that may be found, and local fire and rescue services will need to make their own contacts and arrangements regarding the features of tunnels they cover.

8C2.26 Automatic detection may be fitted, often with smoke or linear heat detection systems or CCTV (Video Incident Detection [VAID]) with monitoring from a control room. Fixed installations may also be fitted, ranging from 'fire points' with hose reels and extinguishers to zoned suppression spray systems.

8C2.27 Often the ventilation systems provided for removal of the car exhaust gases are not suitable for the removal of the products of a fire. Longitudinal fan systems at roof level are generally uni-directional for the removal of exhaust fumes and incorrect use of these fans in the event of a fire may spread the fire products into unaffected areas. One modern example of ventilation is to have large fans placed in a canopy outside the portals which blow through ducts in the roof level and entrain vast quantities of outside air (by the venturi principle) and can create air movement through the tunnel bores. It is vital that local crews and officers are familiar with the ventilation systems on tunnels they cover.

8C2.28 In addition to evacuating the public, many large road tunnels have toll booths to charge users. If these are at the entrances, they can be used to control the traffic access to the tunnels.

8C2.29 From an operational point of view, whether a transport tunnel is traditionally bored or 'cut and cover' construction will have little effect. The main factors will be the length, diameter, profile and installed safety systems. Whether the material above the tunnel is a shopping centre, river, mountain or 1 metre of soil will have little effect on fire development or ventilation, although in some cases the structure of the tunnel may be affected.

8C2.30 It should be considered that the general profile of transport tunnels has the lowest point in the middle. Therefore tunnels may accumulate leaking liquids, vapours, gases, or any applied firefighting media at this point. Many tunnels will have sumps and pump out systems to remove normal road surface water. Fire and rescue services should ensure they are familiar with the capabilities and controls of these systems, and where the liquids may be removed to. As an example this may be known as the 'receptor' in pollution control terms.

8C2.31 Many tunnels will have a plant room for the equipment needed to safely operate the tunnel. These complexes will often have pedestrian access to the tunnel itself, but will be mainly accessed from another road above the tunnel.

8C2.32 In this aerial image of a tunnel on the M25, the control and plant room is indicated by the red box. This complex has pedestrian access into the tunnel bore, but the doors cannot be opened from inside the tunnel. Therefore, access to this complex must be gained from the road network above the tunnel.

8C2.33 It should be considered that some tunnels will have branches underground/underwater (see diagram of Queensway Road Tunnel above). This will make navigating and ventilation difficult and unpredictable and will require crews to be familiar with the layout and have effective information on scene (see operational considerations below).

Aerial view of the Dartford river crossing between Essex and Kent

(For scale, the distance between the portals at each end is 1.5 kilometres)

An example of complicated layout of a road tunnel

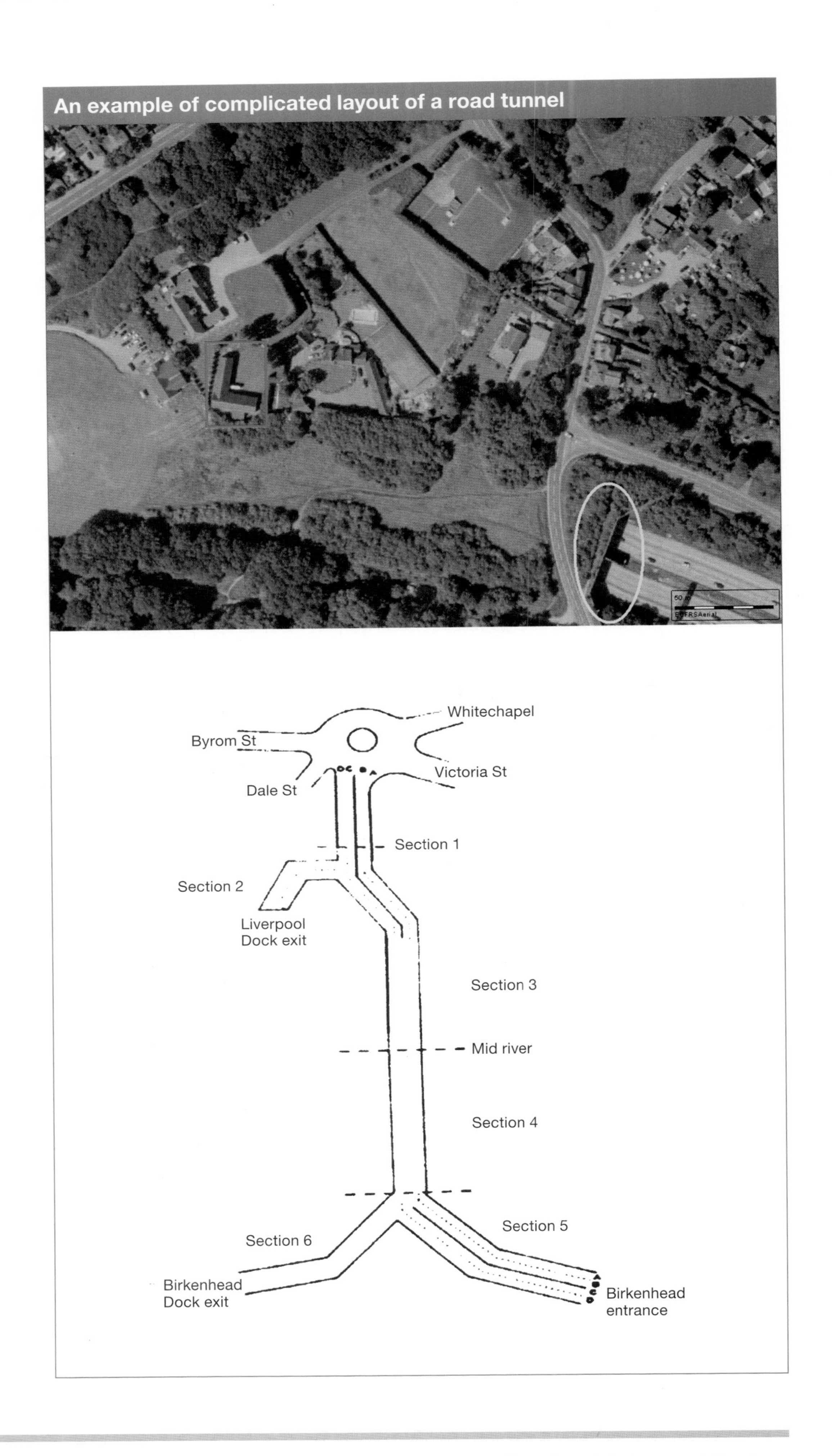

Rail tunnels

8C2.34 Railway systems in tunnels are likely to fall into one of two main categories:

- mainline services running through tunnels
- dedicated below ground train systems.

8C2.35 These may have different design features and pose different operational problems. Mainline services running through tunnels will have track, carriages and power systems which are mainly designed for operation in open air, on the surface. Dedicated train systems will have the problems posed by tunnel operation built into the system's design.

8C2.36 When dealing with a railway tunnel incident, operational issues related to tunnels should be considered in conjunction with design features and operational issues related to railway incidents.

8C2.37 Whichever type of system is in use, some of the features specific to railway tunnels might include:

- characteristics of railway use mean that the tunnels can be longer than other sorts of tunnels
- often large numbers of the public will be on a train, and there may be a lower level of staff supervision in the event of an incident
- other rail vehicles held in tunnels may become hot and uncomfortable causing passengers to become unwell.

8C2.38 Incident Commander's should also give consideration to the following when dealing with railway incidents in tunnels:

- the type of line – single or multi directional
- whether or not the line is electrified (overhead line equipment or third rail)
- whether the overhead line equipment or associated equipment is involved
- water supplies
- emergency lighting
- access to track level and transportation of equipment
- exhaust fumes from petrol driven light pumps, generators and positive pressure ventilation fans.

Dedicated below ground train systems

8C2.39 The best known dedicated systems in the UK are the Channel Tunnel running from Kent to France and the London Underground system.

8C2.40 As the systems are so specific and individual, services with one such system in their area should have standard operating procedures for the specific system and/or site specific risk information cards.

Mainline services running through tunnels

8C2.41 It is estimated that there are approximately 650 railway tunnels in regular service equalling over 200 miles of tunnel in the UK. Most Fire and Rescue Services will have mainline railway tunnels within their area. These will vary in length and complexity and the services may need only a standard operating procedure to cover all the tunnel systems. An example of one of the longest tunnels is described below.

Severn rail tunnel

8C2.42 The Severn Rail Tunnel was built between 1873 and 1886. It is over four miles (7,008 m) long, although only two and a quater miles (3.62 km) of the tunnel are actually under the River Severn. For well over 100 years it was the longest mainline railway tunnel in the UK, until the two major high speed tunnels (London East and West) were opened in 2007 as part of the Channel Tunnel Rail Link.

8C2.43 The tunnel is a critical part of the trunk railway line between southern England and South Wales, and carries an intensive passenger train service, and significant freight traffic. The signalling arrangements are such that almost the whole length of the tunnel is a single signal section, limiting the headway of successive trains.

8C2.44 There is a continuous drainage culvert between the tracks for ground water to flow away to the lowest point of the tunnel, where it is pumped to the surface. The hazard of ignited petroleum running into the culvert in the event of derailment of a tank wagon means that special arrangements have to be made to prevent occupation of the tunnel by passenger trains while hazardous liquid loads are being worked through. Special evacuation arrangements are in place to enable the escape of passengers and staff in the event of serious accident in the tunnel.

Waterway tunnels

8C2.45 Where a navigable canal, river or waterway needs to run below ground level, these can also run through a tunnel. These are different to pipes or mains carrying water.

8C2.46 In the UK, British Waterways operate approximately 50 such tunnels, spread across England, Scotland and Wales. They vary from 50 metres to 5 kilometres in length with varying cross-sectional areas and commonly have masonry or brick construction or linings. Headroom varies from less than 2 metres to over 5 metres above the waterline. The water depth varies, but is commonly 1.2 to 2 metres deep, although there may be deep accumulations of silt.

8C2.47 In general, access to the portals will be good. Only one has any other access points, other than the portals (Standedge Tunnel in Yorkshire shares access points with an adjacent rail tunnel). Many have a towpath for pedestrians and cyclists, although most only have access for the boats.

8C2.48 Ventilation is normally natural ventilation via the portals, although some have air shafts. Only one has a powered ventilation system (Harecastle Tunnel in Staffordshire). Approximately five have lighting systems and only one has automatic fire detection.

8C2.49 The principal sources of major fire loading or hazardous materials are likely to be present in boats using the tunnel, however these can be numerous in some locations and can include powered and unpowered craft.

8C2.50 In addition to the general tunnel incident considerations, waterway tunnels pose additional hazards:

- pre-planning for responders will need to consider procedures for working on, in or near water
- access for crews to an incident will be restricted if there is no towpath
- containment of firefighting water or hazardous materials will be difficult due to the close proximity of the receptor to the source
- rarity of these tunnels mean crews may be unfamiliar with the risks posed and specialist equipment may not be available (e.g. boats for firefighting)
- lack of fixed installations for firefighting or ventilation
- lack of control over access for boats and the public.

Pedestrian and cycle tunnels

8B2.51 These can range from traditional subways or underpasses (where pedestrian and cycle traffic is taken away from the main road network for reasons of safety) to large-scale structures specifically constructed to allow pedestrian and cycle traffic to pass under rivers or between other underground structures.

8B2.52 The design is commonly stairs, a slope or spiral carousel down into the main tunnel. The construction is often as simple as a concrete tube, either rectangular or rounded in cross-section. Lighting is normally provided and barriers to keep cyclists and pedestrians apart. There is unlikely to be any automatic detection or fixed installations for firefighting.

8B2.53 The inherent fire loading of both is likely to be low, with the lights being the only inbuilt source of ignition and few fuels routinely present. Often they are used less outside of peak hours and this may reduce the likelihood of people being involved in a fire, but may increase the possibility of deliberate fire setting.

8B2.54 Access to these tunnels and horizontal ventilation may be easy when it is a simple slope down to the tunnel level. However, where access is gained via a vertical shaft with lifts, stairs or carousel slopes, this will present access and ventilation challenges and some of these tunnels are in excess of 500 metres (e.g. The Woolwich Foot Tunnel beneath the River Thames).

8C2.55 The more complex the access route and the greater length of the tunnel, the more limited fire service radios are likely to be in use. Leaky feeders may be fitted for mobile phone use (as in the Woolwich Tunnel) and individual fire services should check the communications arrangements inside tunnels they are likely to respond to.

Cable and communication tunnels

8C2.56 A cable tunnel is an underground tunnel that carries power or communications cables. The size and type of cable can vary but they form part of the UK's critical national Infrastructure and therefore have significant implications on the community.

8C2.57 These tunnels can be large. The Elstree to St Johns Wood Tunnel in London, for example, is 21km long and 3m in diameter. It is a concrete segment lined tunnel with 6 permanent shafts up to 42m deep with access buildings at ground level. Access shafts will vary in size (St Johns Wood Tunnel shafts are either 7.5m or 10.5m in diameter and up to 42m deep).

Ventilation

8C2.58 Cable tunnels can carry power cables capable of generating considerable amounts of heat, and will usually contain some form of ventilation. This may consist of natural ventilation, assisted by portable motorised fans stored close to the shaft entrances. In longer tunnels this may consist of numerous ventilation shafts, and a range of electrically driven variable speed fans. These fans are often mounted in acoustic enclosures within the shaft in order to reduce noise levels. This ventilation will not necessarily be provided just for smoke control, or be built for fire conditions. The outlets themselves may be a cause of false alarms, or multi-sited attendances to a single incident.

8C2.59 In order to provide a safe means of escape from the tunnel in emergency conditions, some staircases located in the shafts may be pressurised with fresh air supplied from a standby fan system. The internal pressure in these staircases will usually be controlled by pressure relief vents located in the staircase walls to allow effective sealing whilst still being able to open the doors. Motorised ventilation will usually be controlled via a switch room.

Firefighting and fire detection

8C2.60 Some cable tunnel shafts are built with a dry falling main at each shaft to allow fire fighting in the tunnel without personnel having to make an entry.

8C2.61 In high voltage facilities considerations should include the fact that they are normally unstaffed. A central control room may be provided. At protracted incidents a responsible person from the utility company may be requested to attend the incident site (Responsible Person at Silver). Any intervention strategy should consider providing premises information boxes and ensuring appropriate access to any secure site is available, which can include via third parties premises.

8C2.62 It may take time to isolate high voltage electricity and staff must pay close attention to the warning notices and contact numbers provided. Some compartmentation may have been provided, but cables may travel through it so this may only slow, rather than prevent, fire spread.

8C2.63 In extensive tunnel systems, a fire and alarm system will often be installed. This may cover multiple zoned areas covering the shaft base, shaft staircase and switch room. In some cases an early warning smoke detector units may be installed in conjunction with combined smoke/heat sensors. In general, the outbreak and spread of fire within the building is prevented by fireproof cables and shutters.

8C2.64 Some more recent tunnels contain advanced suppression systems such as water misting installations.

Shafts and headhouses

8C2.65 Intermediate shafts and headhouses may be sited at key points along the route of the tunnel. These are unlikely to be obvious and may appear to be ordinary intake rooms. Care is needed when using any lifts provided as they may be for equipment only, or may not be fire fighting lifts.

Other features

8C2.66 Many cable tunnels are fully serviced with lighting and power, however moving around may be difficult as the amount of cable laid and racking provided can be extensive and present significant hazards. 'Wayfinder' facilities may be provided. Blast doors, slam doors (stopping water flow, or charging hoses pushing closed) and dead end conditions may exist.

8C2.67 A leaky feeder may be installed to provide telecommunication facilities.

8C2.68 Because of the enclosed nature of these tunnels, gas monitoring equipment may also be installed at each shaft position.

8C2.69 The intermediate shafts, which are usually capped on completion of construction with a single manhole cover, provide access for personnel. These covers may be secured with special keys or an automatic alarm system to detect any open pit cover. Careful consideration must be taken before committing crews and equipment into tunnels via vertical ladders as they offer no lobby protection or fire separation from any incident.

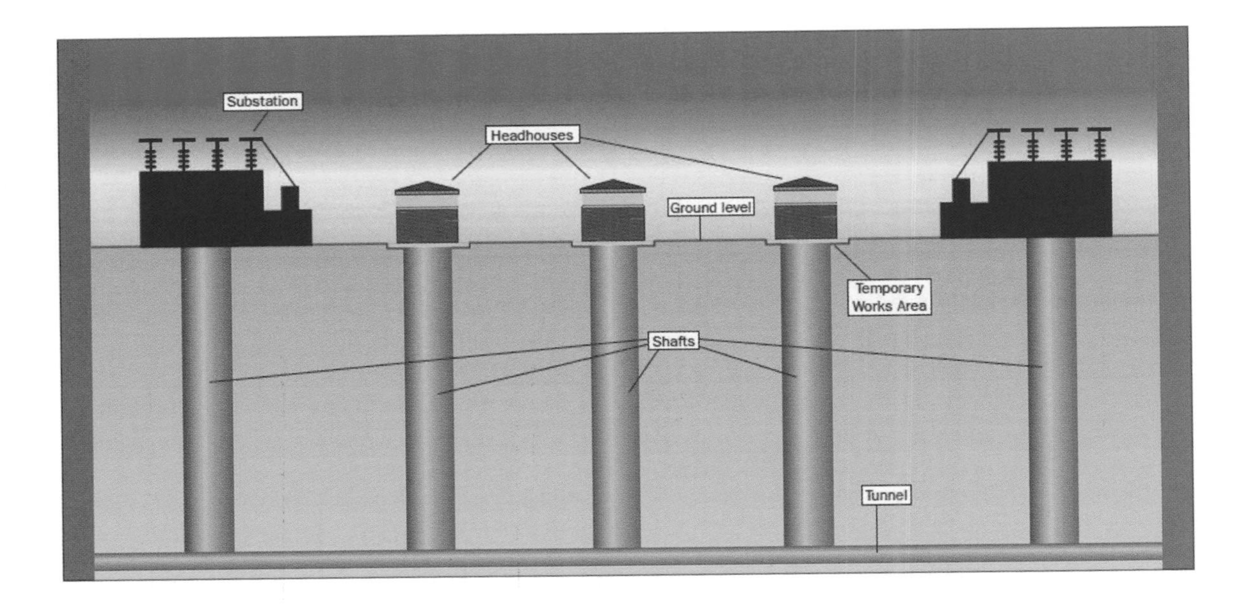

8C2.70 The cables themselves may be contained within formed concrete troughs built into the tunnels. More recent tunnels have seen cables being fixed directly to the tunnel lining and then covered with a cement bound sand mixture, providing a level floor across the tunnel width.

Operational considerations

- cable tunnels can be very complex in layout, deep with dead end passages , over different levels, with difficult access and egress
- until recently the facilities for fire and rescue service purposes have been limited in the design process
- hose lines passing through limited lobby protection doors allowing smoke to spread along infrastructure
- equipment such as lifts may be small and unsuitable for fire and rescue service purposes
- limited water supplies, causing long hose lengths
- old cables may exist that are not 'low-smoke' when involved in fire, creating considerable smoke logging
- cables running through fire stopping, with risk of smouldering cables burning through protection
- several scenes of operations may develop over different fire and rescue service areas
- delay in realising the connection between the separate events, with conflicting tactics creating a risk of harm to responders elsewhere in the infrastructure
- limited telemetry and radio communications

- limited space for moving about, with increased potential for snagging equipment and/or personal protective equipment on cable brackets and associated equipment
- failure of cable brackets and clips due to heat leading to trailing cabling
- power supplies and fixed installation firefighting equipment may be under the control of different organisations, companies or government bodies, causing delay in identification and application of appropriate control measures
- limited or absence of fire ventilation system.

Sewers and water tunnels

Sewers

8C2.71 In most cities and large towns a network of sewers has been established for the discharge underground of domestic sewage, trade wastes and rain or storm water. They are generally described as either local or main sewers. Local sewers are small in diameter and discharge into the larger main sewer system, which in turn carries the waste materials eventually to one of the purification works for ultimate disposal. Normally the system is arranged in such a way that the sewage is moved from the small local sewers to the final discharge point at the purification works by gravitation. In some cases both the sewage and overflow has to be pumped, where insufficient gravitational force is available.

8C2.72 Detailed information on sewage systems can be found within the Fire and Rescue Service Manual titled Environmental Protection.

8C2.73 Within the network of main sewers are:

- main sewers
- intercepting sewers, which relieve the local and main sewers
- storm relief sewers are usually deeper than the others and come into use only when the normal system is so overloaded by excessive rainfall that the sewage rises above the side weirs and overflows into the storm relief sewers.

8C2.74 These sewers form an intricate arrangement of pipes ranging in size from about 150mm diameter to as large as 7 metres in diameter and at a depth of anything from under a metre to approximately 50 metres below ground.

8C2.75 Local sewers feed into the main trunk sewer system and do not normally exceed 2m x 1.5m in size. They have a varied cross section and, in general, are less than 6m below ground level.

8C2.76 The amount of flow in a sewer depends upon:

- the extent of the area it drains
- the gradient of the sewer
- the weather, both in the immediate area and surrounding locality.

Ownership and responsibilities

8C2.77 Water authorities are responsible for the maintenance of sewers and over the years have developed progressively more elaborate safety measures to protect the teams that work in them. Despite these additional measures, sewage workers are occasionally trapped or overcome by the toxic atmosphere sometimes present in sewers and the attendance of the Fire and Rescue Service can be requested to perform a rescue.

Construction features

8C2.78 Many sewer openings are located on or near to roads. Due to the construction of sewers, access is extremely limited and natural lighting is virtually non-existent.

8C2.79 Sewer construction features can be divided into the following categories:

- access points
- inspection covers
- vertical shafts
- chambers.

Access and egress points

8C2.80 Access and egress from sewers will be difficult due to the nature of sewer construction. Access is generally through an inspection cover which will lead into a vertical shaft with access ladders leading down to the main sewer complex some depth below ground. To enable repairs and maintenance to be easily carried out, and to facilitate ventilation, sewer systems are provided with vertical access shafts by which engineers may gain access from the street or footpath level.

Inspection covers

8C2.81 Inspection covers are normally circular and approximately 550mm diameter but are sometimes found triangular or square in shape. They are constructed of heavy cast iron to withstand the weight of traffic, and so that they cannot be removed by one person. Inspection hatch covers vary in size and construction depending upon the position and depth of the sewer served. These may be a two person lift for Fire and Rescue Service personnel, but in some circumstances may require specialist equipment from the water authority to lift access covers.

Vertical access shafts

8C2.82 Vertical shafts can vary in depth from 6 metres to over 30 metres and are provided with either a galvanised iron ladder or step irons fitted internally. These are designed to allow one person to descend the ladder in comfort and therefore space within a circular shaft is normally 700mm across the shoulders and 800mm front to back. Manholes and shafts are designed primarily for the use of maintenance staff carrying only a few tools.

8C2.83 Where the depth of a sewer is more than about 6m below the surface it is normal for the ladder climb to be interrupted by a platform sited halfway down the shaft.

8C2.84 These platforms are usually about 1m square. The ladder from road level descends directly to the platform where as the lower ladder is offset, starting at least 1m above it. A safety bar or chain is fitted at a convenient height above the edge of the platform.

8C2.85 In an alternative arrangement the ladder is continuous and platforms are provided at intervals. The ladder passes through small openings in the platforms, and these openings are fitted within hinged metal gratings.

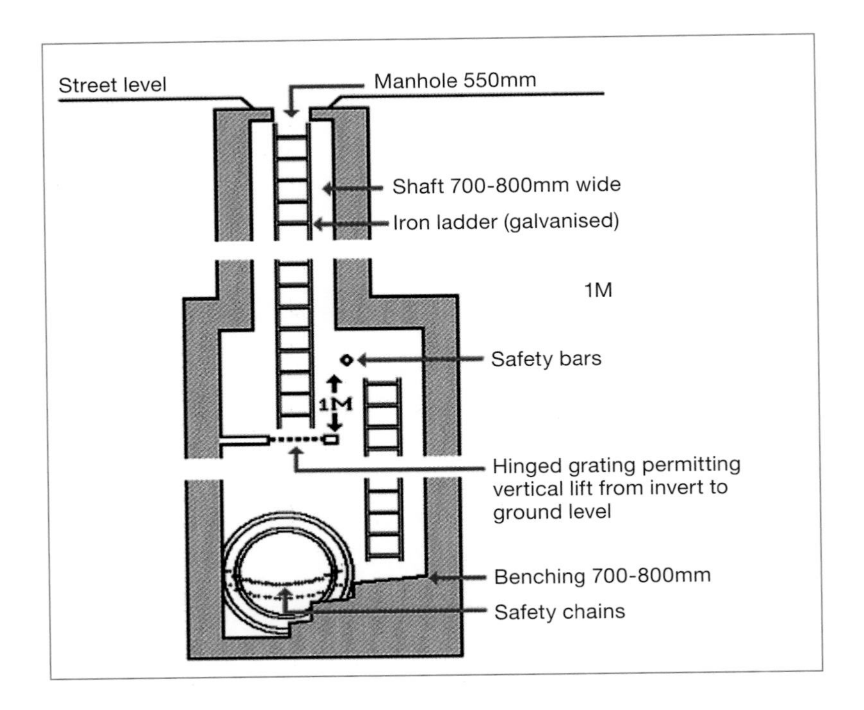

8C2.86 Personnel wearing breathing apparatus will experience considerable restriction to their movements when descending and ascending. Incident commanders will need to consider the amount of time and resource required to get appropriate equipment from the surface to where it is needed.

Chambers

8C2.87 The normal size of a chamber for a small diameter sewer is 1.5m in width. Chamber heights are often up to 2m to enable a person accessing the chamber via the benching to stand upright.

8C2.88 The platform in a sewer chamber slopes towards the sewer invert to facilitate drainage. Generally the angle of the slope is 1 in 6. When wet these platforms present a slip hazard to personnel, especially when wearing rubber boots. In many sewers a safety chain or guard bar is placed across the sewer immediately downstream from the access shaft to act as a handgrip for personnel working within the area.

Internal considerations

8C2.89 Due to the construction of sewers, access is extremely limited and natural lighting is virtually non-existent. Both of these factors increase the risk of injury to personnel when gaining access and moving around the sewer.

8C2.90 Once inside the sewer personnel could potentially be confronted with a complex layout with hazardous conditions and difficult egress which will increase the risk of disorientation and the psychological stress on personnel.

8C2.91 Personnel must consider the potential of the quantity of sewage flowing along any sewer at a given time and will depend on the extent of the area it serves, the time of day and the climatic conditions.

8C2.92 The system of sewage pipes may be arranged in such a way that if work of any kind is carried out in a section, flow may be diverted to alternative routes thus reducing the flow.

Water tunnels

8C2.93 There are very few water tunnels in existence, construction tends to vary, but many are likely to be considered as confined spaces.

8C2.94 The water utility is normally the owner of all such assets, however there are still many un-adopted sewers etc that remain in private/council ownership. It is dangerous to enter modern water tunnels as they are actively transporting often high volumes of water. The flow of water in the tunnel is controlled by a system of sluices and these could be opened at any time filling the tunnel to capacity. Most of the modern water facilities are secured and tunnels are barred at the ends.

2.95 In general modern water facilities are found around the reservoirs and rivers where water collection is active and where pre-existing tunnels have been replaced to allow larger modern roads to be built above.

2.96 The main reservoirs will be found to feature large scale stone water tunnels, towers, embankments and underground chambers that are in use today for water collection and as parts of the reservoir overflow system.

Operational considerations

2.97 Incidents involving Sewers or water tunnels will present serious risks to firefighters:

- Toxic gases such as hydrogen sulphide and flammable gases such as methane, may be present within the sewer network system. Gas detection equipment may be used to detect levels before committing crews, along with upstream and downstream ventilation through removal of manhole covers
- Local authority engineers carry out continuous maintenance, improvement and repair programmes to the sewer system. They are capable of diverting the flow of sewage by the use of sluice gates and penstocks thus reducing the flow
- Personnel must consider the potential of the quantity of sewage flowing along any sewer at a given time and this will depend on the extent of the area it serves, the time of day and the climatic conditions
- For all practical purposes, hand held radios will be severely limited in their use as the frequencies used will only propagate through thin walls and the energy is absorbed quickly by the walls and water within the sewer system. Any equipment, including radio equipment, used should be intrinsically safe

- Normal practice is for someone to be sited in fresh air to monitor work and the safety of workers. They normally work for the sewer company and will be a useful source of information
- Heavy rain, even some distance away, may cause flash flooding or significantly change in the working environment in a sewer tunnel.

Tunnel safety management

General

8C2.98 Under normal operating conditions, the level of management control will depend on the type and the occupancy of the tunnel. Safety management controls could include anything from a well equipped tunnel fully monitored with a local control centre with a high level of resilience for e.g. an urban road tunnel; to a non-monitored utility tunnel that only has controls applied when maintenance work is in progress.

Non–monitored tunnel

8C2.99 Although constant human supervision of the tunnel is not necessary, action requiring human intervention must be taken in the tunnel in the event of an incident or emergency. As a minimum, whenever maintenance or other work is in progress appropriate control measures must be put in place for the safety of persons at work.

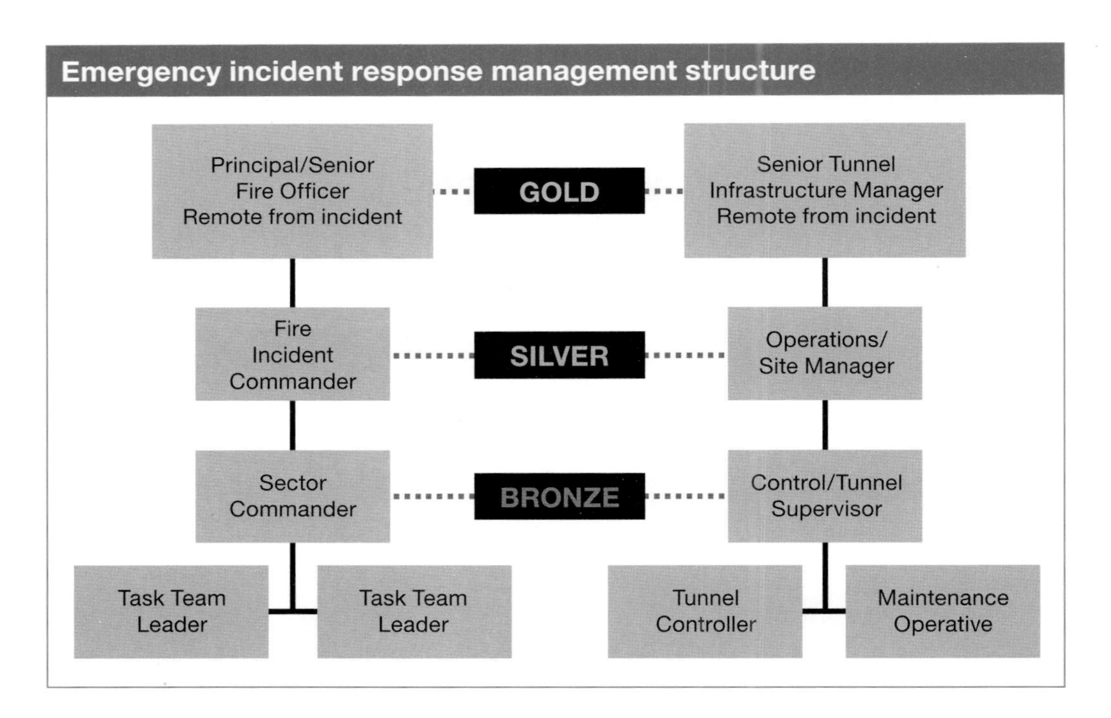

Monitored tunnel

8C2.100 The amount of traffic or the characteristics of the tunnel may make it desirable for there to be constant human supervision. The term "human supervision" refers to the active presence in a monitoring/control station of a person equipped with the means for displaying the interior of the tunnel and its surroundings, for receiving alarms and for initiating the application of appropriate actions to deal with any abnormal situation.

8C2.101 This supervision may be remote and common to several infrastructure units or an entire traffic or rail route.

Local control centres

8C2.102 The local control centre is dedicated to one tunnel (or to a small group of tunnels very close to each other). It is located very close to the tunnel(s). It contains all the necessary equipment to operate the tunnel and all the people working in this control centre are involved in the tunnel operation. Depending on the use of the tunnel and the technology employed, this control centre could be extensive and complex.

Remote control centre

8C2.103 In some cases tunnels can be located on major motorways or on the urban network, when the whole route or network is controlled by a control centre. Very often this sort of control centre is located close to one tunnel but if the route comprises of many tunnels it may be further away. It contains all the necessary equipment to operate the network and the tunnels. This type of arrangement can be very useful on a motorway network or for a group of urban tunnels. For example all the M25 Tunnels are controlled remotely from a control centre at Dartford Tunnel, and the traffic management of those tunnels is managed remotely from the Highways Agency Regional Control Centre at South Mimms in Hertfordshire

Technical management systems – automatic/monitored

8C2.104 Management systems can vary from nothing at all in some sewer and water tunnels to very complex computerised management systems in modern road and rail tunnels.

8C2.105 Where there is a well-equipped tunnel with a local or remote control centre, the facilities, including automatic facilities have to be monitored to some degree by a technical operator, using a wide range of technical data and monitoring equipment.

8C2.106 A technical room in the control centre will provide some or all of the following information:

- location of the incident
- tunnel environment monitoring (air quality, external light levels, weather conditions, etc.)

- mains power supplies monitoring and control
- emergency power supplies availability with automatic and manual override control
- equipment monitoring – functionality, availability, condition, performance (ventilation, pumping, lighting, communications, sensors, etc.) usually automatic with manual override
- safety systems functionality, availability.

8C2.107 A technical operator can have at their disposal the following facilities:

- alarm systems
- communication and recording systems (closed circuit televisions capture, video incident detection)
- supervisory control and data acquisition (SCADA) system (computer driven data collection, processing and automatic control system).

8C2.108 The above equipment, properly designed and integrated, should be capable of automatic management of the tunnel technical equipment in all normal conditions. A sophisticated supervisory control and data acquisition system can also be pre-programmed to deal automatically with many abnormal emergency situations.

8C2.109 The main role of the technical operator is the total overview of the equipment to ensure its correct function and to respond correctly; if necessary with manual intervention in case of malfunction or emergency operating conditions.

System monitoring

8C2.110 The technical operator is also responsible for the management of the maintenance response in case of equipment fault or breakdown and can be responsible for:

- analysing the technical malfunctions (problems in power supply, communication networks, facility breakdowns)
- assessing the safety margins in terms of technical operation with eficient equipment
- managing the actions on these facilities.

Manual controls

8C2.111 As the name implies where complex monitoring and control systems are not in place, there will be a reliance on manual control of tunnel systems to raise the alarm in the event of an incident and operate systems, this can include:

- manual observation of CCTV
- physical switching of key systems into incident mode:
  - traffic control signals
  - lighting
  - ventilation
  - communication systems.

PART C–3

Underground structures

General

8C3.1 As the demand for space increases, new developments are increasingly looking to use underground facilities in order to:

- reduce disruption to existing infrastructure
- reduce the environmental impact of commercial undertaking
- increase the communications and utilities available to built up areas
- take advantage of the relative security provided by sub-surface developments.

8C3.2 Consequently it is possible that fire and rescue services will plan and respond to underground incidents involving the full range of building uses. The following provides an insight to demonstrate the range of significant uses that underground locations are being put to:

Underground railway stations

8C3.3 These can range from quite simple structures to major transport interchanges, with complex and specially engineered solutions for emergency management and response.

8C3.4 The management of underground stations will vary at different times of the day depending on whether they are occupied or not. During public opening hours underground stations have management present and are normally unstaffed at all other times.

8C3.5 When responding to these locations the site specific operational considerations may include:

- responding to the agreed rendezvous points and gathering information from plans boxes and a member of staff assigned to meet and brief the Incident Commander
- use the managed facilities, for example station ventilation or compartmentation to assist with initial reconnaissance
- deploy crews with appropriate resources to initiate intervention
- identify correct water supply and falling mains equipment, this can be complicated in older, large or redeveloped locations

- initiate and maintain communications with crews
- the Incident Commander is to identify any limitations on the range of communications from the station.

Underground storage facilities

8C3.6 These facilities will usually be a location that has been converted over from a previous use, normally a mine. The materials that can be stored can vary from archiving often of significant national value to commercial wine storage. These facilities can be complex and, depending on the materials stored and arrangements for fire protection and separation in place, can present fire and rescue service responders with a challenging environment. At these locations operational considerations for Incident Commanders will include:

- making contact with on site personnel and interrogating the status of any fixed fire fighting installations
- identify the progress of staff and customer evacuation
- identifying the likely spread of the incident and develop a realistic intervention plan
- attempt to hold the spread of the fire until concerted attack can be established
- consideration of implementing fire break if resources and conditions permit
- consider specialist advice or equipment from, for example, mines rescue teams.

Underground wastewater treatment plants

8C3.7 For environmental and community reasons major waste water treatment developments are being located below the surface. This may have the advantage of making the environmental impact of a hazmat release more manageable, Additional operational considerations at these locations will include:

- obtaining specialist advice from on-site personnel as to the hazards and risks associated with any treatment or processes underway
- implement appropriate controls to protect personnel from potential hazards
- consider the wider environmental or community implications when developing the risk assessment and determining the appropriate response
- consider any off-site planning arrangements and facilities that the Fire and Rescue Service can provide to assist in containment or protection from pollution.

Ministry of Defence installations

8C3.8 The Ministry of Defence use a range of facilities, which includes either purpose built or converted over for defence purposes. The use the infrastructure is put to can range from administrative to storage. Any use could involve sensitive purposes with high strategic military importance. It is imperative that the Fire and Rescue

Service look to engage the local Ministry of Defence representative, at an appropriate level within the Ministry of Defence, to ensure that intervention plans are appropriate.

8C3.9 Some Ministry of Defence establishments with a strategic underground risk are provided with a professional on-site fire and rescue service to allow rapid intervention and thereby increase the protection of Ministry of Defence strategic capabilities. In these cases it is essential that Ministry of Defence and local authority fire and rescue services work together in the pre-planning and implementing of arrangements for firefighting and rescue in Ministry of Defence facilities. Integrated operational response plans should be regularly tested.

Mines

3.10 Incidents in these locations are not normally responded to in the same way as incidents in other infrastructure. This is because of legislative differences that are discussed further in Appendix 2.

PART C–4

Features of tunnels and underground structures

General

8C4.1 The types of infrastructure covered by this guidance may have a range of features and facilities installed that are specifically designed to assist operational personnel during Fire and Rescue Service response. It should be noted however that many shorter tunnels will have few of these features and facilities available. The extent and type of installed facilities in longer transport tunnels and other underground infrastructures will reflect the risks involved, the age and use of the infrastructure along with any statutory requirements placed upon the owner or infrastructure manager. Included in this will be the quality of management, and the installation and maintenance of any fixed systems.

8C4.2 The most effective safety features will be those that are developed and installed following discussion with the Fire and Rescue Service, which look to ensure effective evacuation procedures and support meaningful Fire and Rescue Service intervention. When planning or responding it should not be assumed that all parts of the infrastructure will be protected by the same features to the same standard.

8C4.3 It should be noted that various parts of any infrastructure may be under the control of different organisations, where the provision of equipment may vary depending on the interpretation of risk. This may be particularly true in large civil works involving a number of contractors.

Ventilation

General

8C4.4 The purpose of ventilation is normally for keeping the users of infrastructure comfortable, equipment cool, or assisting with the control of hot gasses or fumes. For the purposes of Fire and Rescue Service operations ventilation systems can be divided into two principal types:

- natural, caused by the flow of air through the infrastructure's openings
- mechanical, where a ventilation system serves a specific function. For example cooling, removing fumes, controlling smoke or fire.

8C4.5 It should be noted that ventilation systems are more applicable to highway tunnels due to high concentration of contaminants. Rail transit tunnels may have ventilation systems in the stations or at intermediate fan shafts, but during normal operations rely mainly on the piston effect of the train pushing air through the tunnel to remove stagnant air. Some rail transit tunnels have emergency mechanical ventilation that only works in the event of a fire.

Natural ventilation

8C4.6 This type of ventilation is provided by a number of means including:

- atmospheric conditions
- vehicle movement (including train 'piston effect')
- structural features (including former steam chimney vents in old rail tunnels, or vertical shafts in mines).

8C4.7 A naturally ventilated tunnel is as simple as the name implies. The movement of air is controlled by meteorological conditions and the piston effect created by moving traffic pushing the stale air through the tunnel. This effect is minimized when bi-directional traffic is present. The meteorological conditions include elevation and temperature differences between the two portals, and wind blowing into the tunnel. Picture 1 shows a typical profile of a naturally ventilated tunnel. Another configuration would be to add a centre shaft that allows for one more portal by which air can enter or exit the tunnel. Many naturally ventilated tunnels over 180 m (600 ft) in length have mechanical fans installed for use during a fire emergency.

Picture 1: Naturally ventilated tunnel

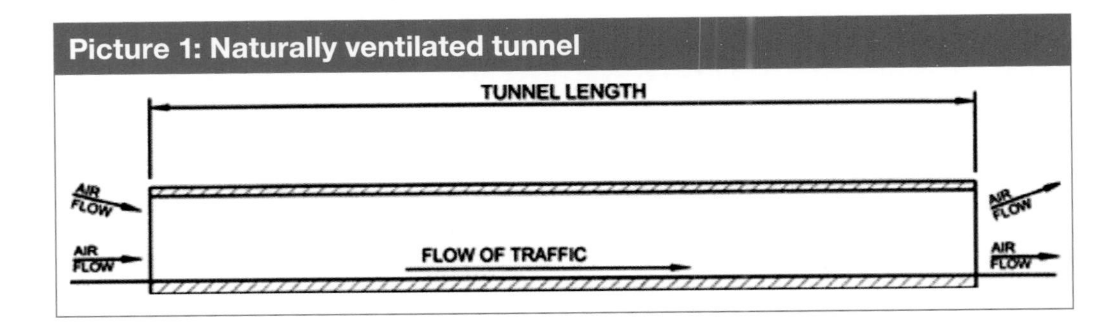

Mechanical ventilation system types

8C4.8 Tunnel ventilation systems can be categorized into four main types or can be found in any combination of these four or combined with natural systems.

8C4.9 The four types are as follows:

- longitudinal ventilation
- semi-transverse ventilation
- full-transverse ventilation
- single-point extraction.

LONGITUDINAL VENTILATION

8C4.10 Longitudinal ventilation is similar to natural ventilation with the addition of mechanical fans, either in the portal buildings, the centre shaft, or mounted inside the tunnel. Longitudinal ventilation is often used inside rectangular-shaped tunnels that do not have the extra space above the ceiling or below the roadway for ductwork. Also, shorter circular tunnels may use the longitudinal system since there is less air to replace; therefore, the need for even distribution of air through ductwork is not necessary. The fans can be reversible and are used to move air into or, in the event of a fire, out of the tunnel from both portals. Picture 2 shows two different configurations of longitudinally ventilated tunnels.

Picture 2: Longitudinally vented tunnels

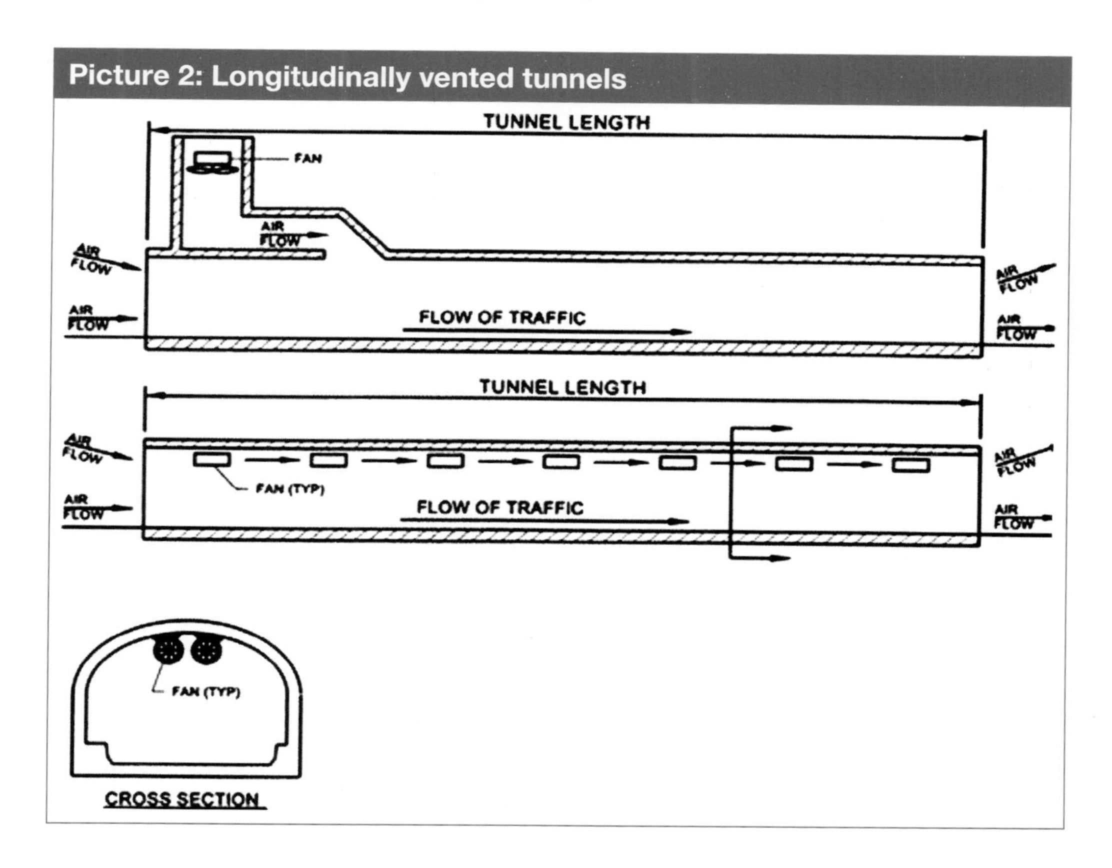

SEMI-TRANSVERSE VENTILATION

8C4.11 Semi-transverse ventilation also makes use of mechanical fans for movement of air, but it does not use the roadway envelope itself as the ductwork. A separate plenum or ductwork is added either above or below the tunnel with flues that allow for uniform distribution of air into or out of the tunnel. This plenum or ductwork is typically located above a suspended ceiling or below a structural slab within a tunnel with a circular cross-section. Picture 3 shows one example of a supply-air semi-transverse system and one example of an exhaust-air semi-transverse system. It should be noted that there are many variations of a semi-transverse system. One such variation would be for half the tunnel to have a supply-air system and the other half an exhaust-air system. Another variation is to have supply-air fans housed at both ends of the plenum that push air directly into the plenum, towards the centre of the tunnel. One last variation is to have a system that can either be exhaust-air or supply-air by utilizing reversible fans or a louver system in the ductwork that can change the direction of the air. In all cases, air either enters or leaves at both ends of the tunnel (bi-directional traffic flow) or on one end only (uni-directional traffic flow). In the event of a fire the fan can draw air into the tunnel and push smoke or fumes out of the portals.

Picture 3: Supply-air semi-transverse system

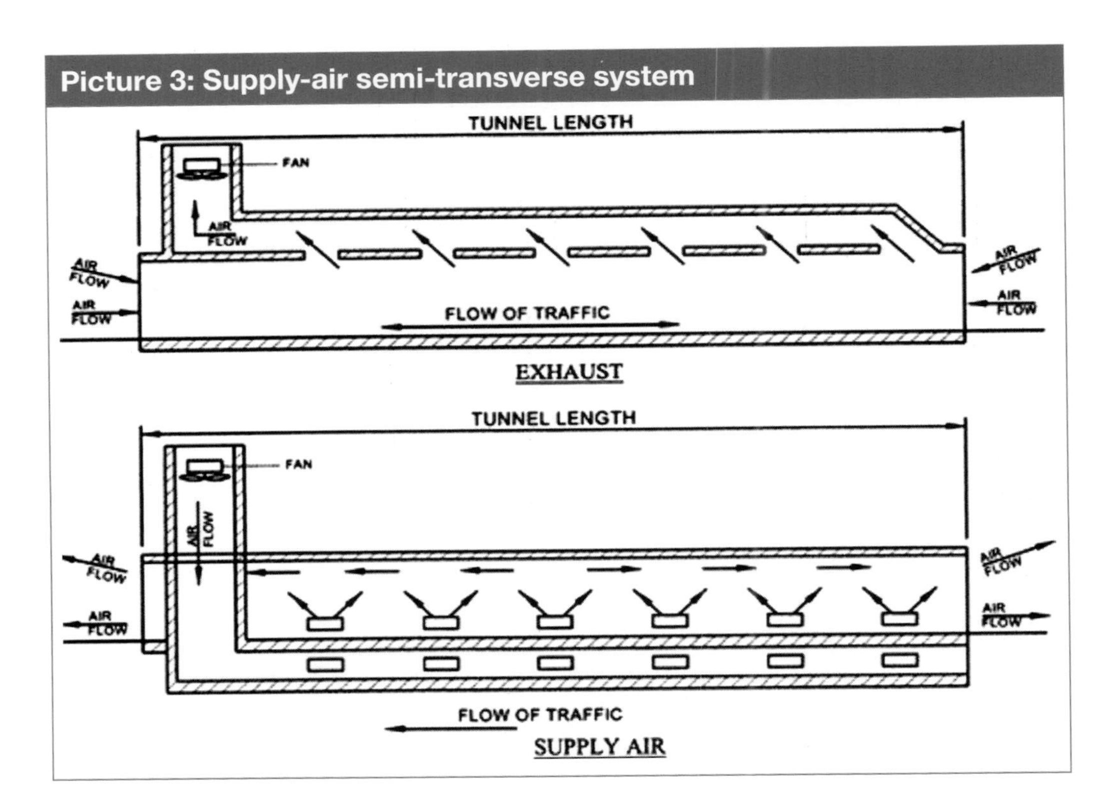

FULL-TRANSVERSE VENTILATION

8C4.12 Full-transverse ventilation uses the same components as semi-transverse ventilation, but it incorporates supply air and exhaust air together over the same length of tunnel. This method is used primarily for longer tunnels that have large amounts of air that need to be replaced or for heavily travelled tunnels that produce high levels of contaminants. The presence of supply and exhaust ducts allows for a pressure difference between the roadway and the ceiling; therefore, the air flows transverse to the tunnel length and is circulated more frequently. This system may also incorporate supply or exhaust ductwork along both sides of the tunnel instead of at the top and bottom. Picture 4 shows an example of a full-transverse ventilation system.

SINGLE- POINT EXTRACTION

8C4.13 In conjunction with semi- and full-transverse ventilation systems, single-point extraction can be used to increase the airflow potential in the event of a fire in the tunnel. The system works by allowing the opening size of select exhaust flues to increase during an emergency. This can be done by mechanically opening louvers or by constructing portions of the ceiling out of material that would go from a solid to a gas during a fire, thus providing a larger opening. Both of these methods are rather costly and thus are seldom used. Newer tunnels achieve equal results simply by providing larger extraction ports at given intervals that are connected to the fans through the ductwork.

Picture 4: Example of a full-transverse ventilation system

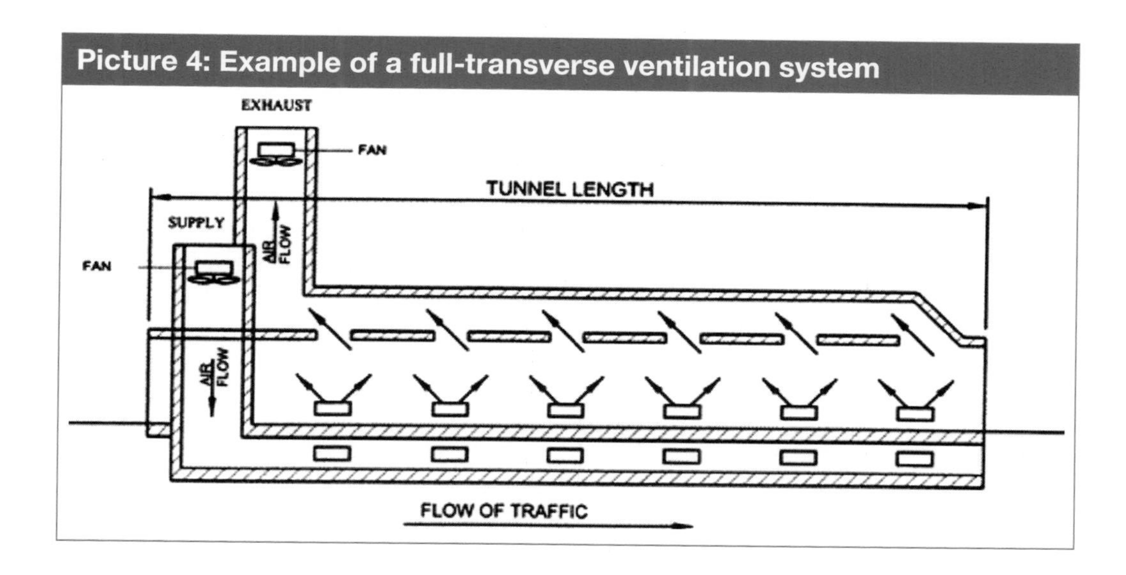

Operational considerations

8C4.14 If the ventilation is mechanical it should normally be known whether it has been provided, designed and rated for Fire and Rescue Service purposes. If it has been provided to control smoke or fumes in an emergency then the Incident Commander can have confidence that it will provide adequate protection to those evacuating the premises, under the guidance of the infrastructure manager, and to responders being committed.

8C4.15 Where mechanical ventilation is unidirectional, or where pre-prepared plans define a default air flow direction, it is important to recognise that this necessarily creates a preferred entry point at the inlet end of the tunnel. Plans should then have responders committed in this direction.

Fire ventilation and smoke control

8C4.16 This is normally a mechanical system designed in consultation with the Fire and Rescue Service to an agreed standard, reflecting the nature and risk presented by the type of infrastructure. It is provided and maintained by the Infrastructure Manager to improve the conditions inside a tunnel or underground environment to allow evacuation and intervention. This type of ventilation is normally operated at the detection of an incident, and is a fundamental aspect of any intervention or evacuation strategy.

8C4.17 Some examples of how ventilation protects responders and the public are shown below. The examples given are for road tunnels, but the principles apply across a range of infrastructure uses.

8C4.18 Smoke control can also be applied to protect shafts and passageways. When planning any system it is important to set robust performance capabilities. These capabilities will reflect, among other things:

- the design size fire
- the type and quantity of combustibles

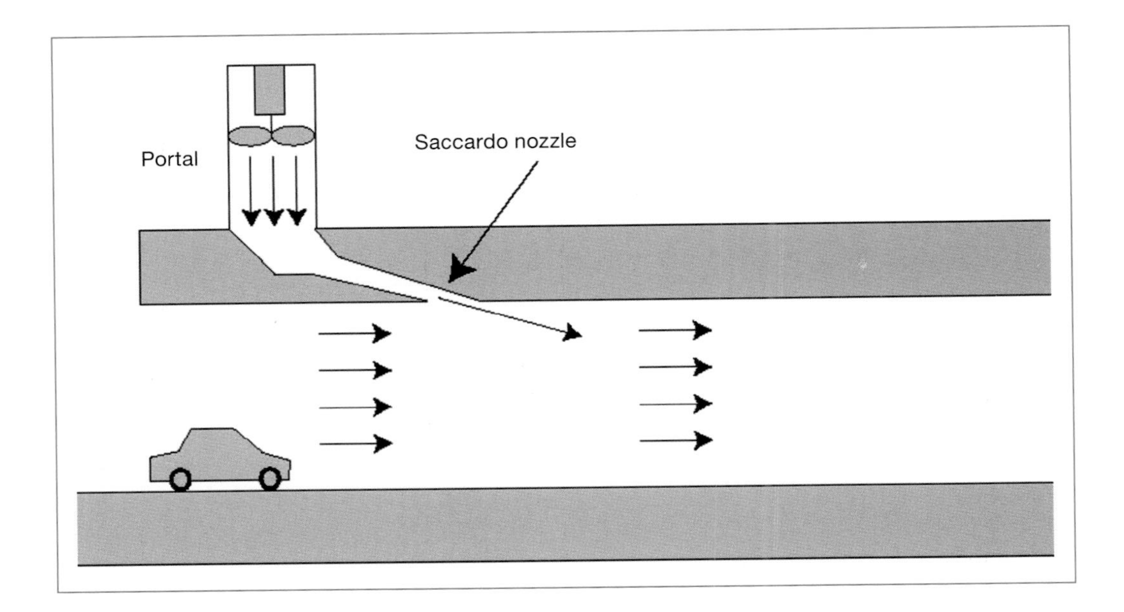

- the evacuation strategy
- the construction, dimensions and use of the infrastructure
- the time available for members of the public/employees to leave the area
- the effect of evacuation and intervention (the number of doors being held open, reducing pressure or interrupting air flow) on smoke free areas
- the time required to evacuate the public
- maintaining a tenable environment to commence meaningful Fire and Rescue Service operations.

8C4.19 The design, extent and responsibilities regarding ventilation facilities that may be available to an incident commander will normally be determined at the pre-planning stage, either as part of the construction or refurbishment stage or as part of the development of appropriate intervention strategies for existing legacy tunnels or other underground structures.

8C4.20 The ventilation in place at an incident can assist the intervention of responders and evacuation of people or, if uncontrolled or inappropriately used, can cause a significant deterioration in the conditions. It is therefore desirable that Incident Commanders have both a general understanding of the ventilation systems and its purpose, along with access to local knowledge about the specific systems that may be in place. This will inform risk assessments and allow for the development and implementation of appropriate operational tactics.

8C4.21 An Incident Commander can normally expect that upon arrival any fire mechanical fire ventilation system will be operating, but a check should be made with the infrastructure manager as to its status. Normally it would be good practice to continue operations with the ventilation system being unchanged. This will reduce the risk of compromising escape routes of persons evacuating.

8C4.22 If the Incident Commander is minded to make changes to any ventilation careful thought of the implications to those directly involved, responders and others in the wider infrastructure should be considered.

8C4.23 Some fire and rescue services have developed ways to augment on site ventilation by providing additional methods of moving or controlling smoke at incidents. This has included the use of appropriate Fire and Rescue Service positive and negative pressure ventilation equipment.

8C4.24 Ventilation in underground fires is dealt with further in Fire Service Manual *Volume 2 Fire service operations – Compartment fires and tactical ventilation*. Chapter 10 basements, underground structures and tunnels.

Emergency access

General

4.25 The type and quality of access arrangements for fire and rescue services can vary greatly dependent on the use, age and location of any tunnel or underground structure. However in many instances the access facilities will be constructed to provide a method of accessing and moving about the infrastructure with Fire and Rescue Service needs embedded in the design, which may include for example:

- hard standing areas for emergency vehicles
- dedicated rendezvous points for emergency vehicles
- Fire and Rescue Service communications extended to cover the rendezvous points, shafts, underground area or tunnels
- premises information boxes or security standard boxes with plans, entry codes or keys
- fire fighting water supplies
- security doors providing agreed method of Fire and Rescue Service entry without unreasonable delay, normally using keys, or entry codes or remote door release devices
- firefighting stairs and lifts
- ventilation system to protect fire fighting access from contamination
- firefighting mains throughout
- firefighting lobbies
- 'Through wall' breechings, to save hose being run through doors
- raised walkways.

8C4.26 Gaining access to the affected infrastructure in a safe and controlled way is a critical part of Fire and Rescue Service operations. The means by which responders may access the infrastructure will vary depending on type, age and location, however the following two categories cover many of the access locations that fire and rescue services are likely to encounter.

Emergency response locations

8C4.27 Emergency response location is the term used in infrastructure where a fire engineered solution has been developed combining the evacuation and intervention strategies with system controls and the structural design into an integrated approach to emergency response. This is normally developed from the planning stage of a project.

8C4.28 These are usually locations that have been designed to facilitate Fire and Rescue Service emergency intervention purposes. (This may not be their exclusive use). These locations are normally agreed between the Fire and Rescue Service and Infrastructure Managers as part of any strategy for emergency intervention.

Intervention point

8C4.29 These are usually locations that are not specifically designed for the purposes of Fire and Rescue Service emergency intervention, but have been adopted as a convenient method of access, particularly on older infrastructure. These can include disused rail stations, tunnel portals and engineering access stairs.

Cross passages (also known as 'cross connections' or 'crossovers')

8C4.30 For infrastructure involving twin bore tunnels there may be means of travelling between bores. These can assist in the evacuation of the public or allowing crews to gain access from an area of relative safety, ie the unaffected bore. The range of cross passages available include:

- a simple arch, for example in older rail tunnels
- a separating door, for example in quite modern road tunnels for intervention purposes or means of escape
- a fully protected passage, designed for Fire and Rescue Service intervention purposes and providing a means of escape, occasionally 20 metres long.

8C4.31 The provision of cross passages allow for flexibility in the Incident Commanders plan. This is critical where the incident has developed in an unconventional or unexpected way, requiring improvisation or adaptation of pre determined intervention strategies.

Twin tube bored tunnel

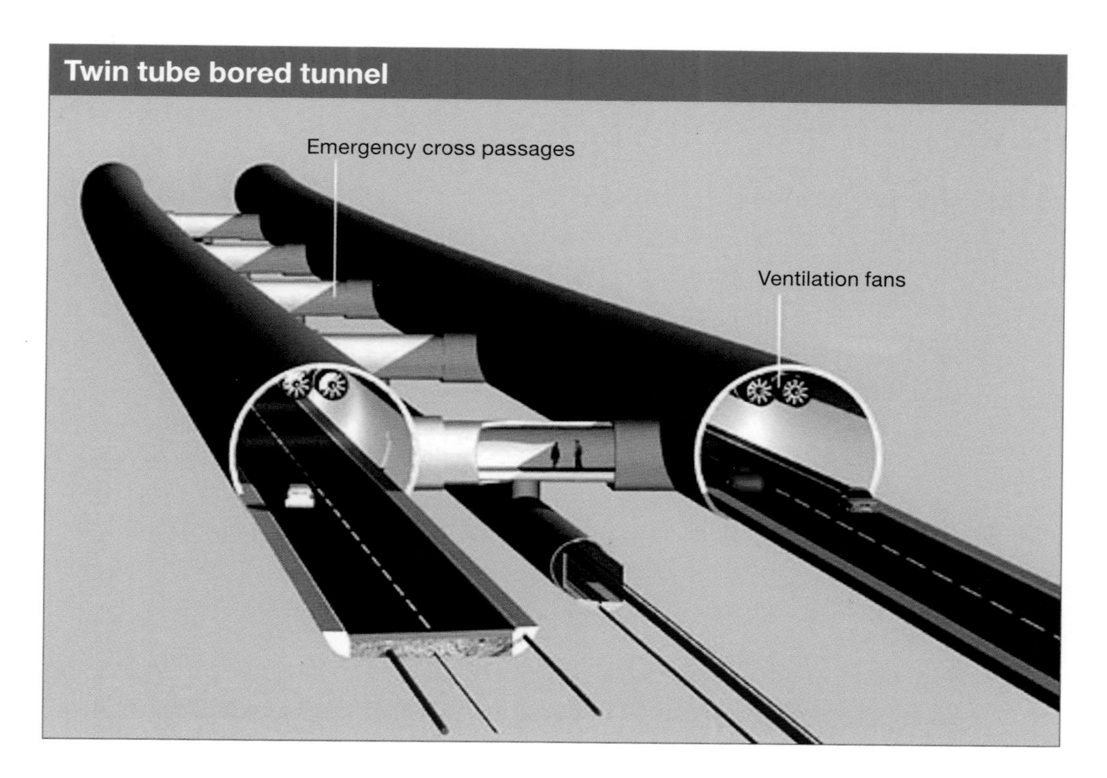

Shafts

8C4.32 In locations where tunnels are deep, it will be usual to provide proper means of escape which will usually involve an escape shaft and sometimes escape passages to the shaft bottom. Often shafts are multipurpose containing services stairways, fans and ventilation ducts or the shaft itself may be the duct. Modern facilities have separation between the escape route and the prevailing conditions in the facility. Shafts will vary in diameter from an access shaft into a manhole of perhaps 1metre diameter up to major escape and ventilation shafts of more than 11m metres diameter.

Refuges (also known as niches)

8C4.33 In tunnels where a pedestrian could be at risk from rail or road traffic, there must be a provision to allow refuge within a short distance. This refuge is either provided by a cross connection or by a niche in the tunnel wall. Cross connections are usually no closer than 250-350 metres apart but niches might be 20m apart.

8C4.34 Niches are usually a recess in the tunnel sufficient to accommodate up to four people.

Automatic fire detection

8C4.35 The provision of automatic fire detection will depend on the age and use of the infrastructure, as well as the owner or infrastructure manager's fire management systems. Care is required when attending incidents as it should not be assumed that all parts of the infrastructure will have fire detection coverage, for example, some transport tunnel access shafts may have its own detection equipment but the tunnels or wider network the shaft connects to may have no automatic fire detection protection at all. Caution should be used when consulting any fire indicator panel, as an incident may be developing, albeit not in the area protected.

8C4.36 If extensive automatic fire detection is provided, depending on the type of detection system, the indicator panel may give an indication on the extent of fire or smoke spread.

8C4.37 Several modern systems incorporate CCTV and fire detection technology. This equipment can analyse CCTV images to detect unusual activity, or initial indicators of fire and bring it to the attention of infrastructure managers.

Lighting

8C4.38 The standard and extent of emergency lighting provided will vary depending on the type, use and history of the infrastructure. This may include times where no emergency lighting is available. Pre-planning and consultation with the infrastructure management should be provided to the appropriate level

8C4.39 as indicated in standard BS 5266. Consideration should be given to the likely protracted nature of an incident when determining if any additional battery capacity should be provided to assist the operational response

Fixed installation

8C4.40 The range of equipment available for Fire and Rescue Service purposes broadly reflects that available elsewhere in the built infrastructure. It is essential to identify if any systems are available and, if automated or manually activated by the infrastructure managers, determine if they have operated. Examples of fixed installation systems that might be built into infrastructure or equipment used underground includes:

- water mist/fog
- sprinklers
- water curtain
- gas suppression/flooding systems
- some very sensitive locations, or vehicles travelling through some types of tunnel, may still retain fixed halon extinguishing equipment.

Water supplies/hydrants

8C4.41 The provision of water supplies for firefighting purposes in the infrastructure ranges from the complete absence of any facilities through to full firefighting main provided intended to serve all areas of the infrastructure. The absence of facilities can be due to the age or use of the location, for example:

- historical infrastructure that did not normally provide facilities
- high voltage electrical tunnels not providing any firefighting facilities, but have recently started to provide mains to access shaft areas only.

Communications

8C4.42 Communications used for Fire and Rescue Service purposes at an incident will be greatly enhanced by pre-planning and testing of the range and extent of signals, including joint testing with other agencies and infrastructure managers.

8C4.43 When determining the type and extent of communications to be requested from the infrastructure managers, the normal Fire and Rescue Service safe systems of work should be reflected. It should be remembered that communications are fundamental to supporting safe systems of work, and any lack of facilities may have a negative impact on operations. Infrastructure managers should understand that without effective communication between all responders the ability to protect people and infrastructure may be negatively affected.

8C4.44 In some circumstances limited installed infrastructure can support adequate communications with existing Fire and Rescue Service systems. For example a relatively short tunnel that allows good, uninterrupted radio coverage in all reasonably foreseeable circumstances should not require the additional provision of a Fire and Rescue Service leaky feeder.

8C4.45 Alternatively even a tunnel during the construction stage may benefit from the provision of enhanced communication facilities that support Fire and Rescue Service safe systems of work, up to and including leaky feeder coverage.

8C4.46 Care is required on the part of the Fire and Rescue Service when accepting the use of the infrastructure's own communication equipment. Particularly if this is hard wired telephone communications to a control point. The potential implications, including the loss of communication at a critical time, require careful examination on the part of the service.

Compartmentation

8C4.47 Compartmentation that would normally apply to the wider built environment may be impractical due to the use of the infrastructure, for example underground railway systems and mines. Consequently many large developments may be provided with a dis-application of building regulations. The local Fire and Rescue Service should look to ensure that any engineered solution and associated intervention and evacuation strategies are robust, compatible and meets the general principles laid out in the Building Regulations.

Plans boxes

8C4.48 There are a number of types of information boxes available to provide Incident Commanders with information upon arrival.

8C4.49 These boxes should be placed in locations where they are immediately visible to first responders arriving at a rendezvous points. They should normally be located in the open air, secured to the structure to which they concern. The purpose of the box should be immediately apparent, as being for fire and rescue service purposes.

8C4.50 If the information contained within the box is sensitive there are a number of security boxes available with limited 'master key' functionality. Here the Fire and Rescue Service has entered into agreement with the infrastructure managers and suppliers, to carry a single key that will open a range of programmed locks, but the local infrastructure manager's key will only open those boxes within their responsibility.

8C4.51 At locations where the contents are less sensitive a box fitted with a standard fire brigade access key is an option, or a padlock that can be compromised using Fire and rescue Service front line equipment.

8C4.52 Plans boxes may contain the following items:

- appropriate plans of the infrastructure as far as any adjacent intervention point
- notification of any temporary unusual risk relating to the site
- access keys
- entry lock combination numbers.

8C4.53 This is an indicative list. The main principle being that the information available for the first responder is uncomplicated, easy to digest and contains the information necessary to inform an initial risk assessment.

8C4.54 Information such as the infrastructure manager's emergency plan may be too cumbersome for the initial stages of an incident's response.

Gas detection

8C4.55 The provision of detection equipment to detect contamination by harmful gasses may be found where the infrastructure manager has assessed there is a risk to employees or the public. These systems may be either fixed or carried in by workers.

8C4.56 Detection equipment will only be enabled to detect a limited range of gasses, normally consisting of those identified in the infrastructure manager's own risk assessment as likely to present a hazard to people.

8C4.57 Care should be taken to ensure gas detection equipment actuating is not confused with smoke automatic fire detection actuation, resulting in an inappropriate investigation being instigated.

Environmental and pollution control systems

8C4.58 Modern infrastructure will have a range of methods for collecting water run off from their normal operations. Where there is a possibility of contaminating water systems, or where the infrastructure cannot move the water to the main sewer system interceptors may be provided.

8C4.59 For example, most longer road tunnels have rainwater runoff drains that are fitted with interceptors because of the risk of fuel/oil leaks. Fire and Rescue Service operations may overload such systems with the potential for flushing contaminants past the interceptors. Pre-planning work should highlight this, allowing the Incident Commander to make informed decisions about collection/containment.

8C4.60 Conversely in older locations and mines, there may be no facility for collecting water, and this may be allowed to collect, or flow to the open, or sink to the water table.

8C4.61 Incident Commanders will need to consider the hazards posed to the environment from any incident or from Fire and Rescue Service actions, and to the collection and containment of hazardous liquids where this is required.

Traffic control systems

8C4.62 Any vehicular movements, rail or road, will need to be managed in order to prevent a worsening of the incident and to protect:

- the incident scene
- vehicles moving through the infrastructure
- responders
- the evacuation strategy.

8C4.63 Control of traffic will normally be carried out using standard methods appropriate for the particular transport type. Some older tunnels may require local plans to be developed to ensure that standard procedures are compatible with the design and facilities provided.

8C4.64 Increasingly for road tunnels the use of barriers or traffic lights are being installed outside the entrance to tunnels. These are operated by the infrastructure managers.

8C4.65 Incident Commanders should be aware that the presence of traffic control systems will not guarantee the compliance of members of the public with those systems when faced with fire or other emergency situations.

Public information systems

8C4.66 The behaviour of members of the public in tunnels can also be managed by the use of public address systems, passenger information media and, for road tunnels via, trunked radio re-broadcasting system. These methods will all be under the control of the infrastructure managers.

PART C–5
Intervention strategies

General

8C5.1 Pre determined intervention strategies for different types of tunnels and underground structures will be determined according to local circumstances relating to resources and risk.

8C5.2 A key element of Fire and Rescue Service intervention strategies at tunnels or underground locations is that the infrastructure manager will follow a number of basic principles:

- The infrastructure should be managed to ensure that all reasonable measures are taken to prevent the incident occurring, and if an incident occurs that reasonable attempts are made to mitigate the effects, for example keep a train moving until it reaches open air or stop adding to any potential fire loading in the area as soon as possible
- The situation is not made worse, for example people and traffic are prevented from entering the infrastructure and anyone not immediately affected can safely make their way out. This will reduce the life risk and reduce the potential fuel loading of an incident
- If the incident does remain within the infrastructure it is positioned as closely as reasonable possible to the emergency response locations/intervention points
- If the incident cannot be removed from the infrastructure the Infrastructure Managers summon the Fire and Rescue Service to respond to the nearest intervention point or emergency response location, normally for tunnels this is to either side of the incident
- If, for whatever reason, the incident occurs elsewhere there are reasonably suitable facilities to allow a meaningful intervention.

Generic underground structures

8C5.3 Underground structures are entirely variable in size and complexity and intervention strategies will be developed by the local Fire and Rescue Service that deal specifically with the circumstances, risks and resources relevant to those locations. However and rescue services may also wish to consider the development of generic policies for sub-surface working, which may include consideration of the following issues:

- the implementation of special procedures at extensive sub surface premises where there is no evidence of unsafe conditions at surface level such as:

- a reconnaissance team wearing breathing apparatus that is not started up to be committed
  - breathing apparatus entry control point(s) to be set up at a bridgehead within the premises.

- Developing a simple method of recording personnel deployed into the infrastructure, appropriate to the hazards
- The use of bridgeheads below ground to enable an incident to be dealt with by establishing a control point with suitable resources and emergency provision. This should be in a safe area as close as practical to the scene of operations making best use of suitable aspects of the infrastructure, and avoiding dead end conditions
- The use of suitably equipped and led reconnaissance teams to investigate the situation and deal with minor incidents.

8C5.4 Some extracts from specific and general intervention strategies are as follows:

Channel Tunnel Rail Link (also known as High Speed 1)

8C5.5 This system is an extension of the Channel Tunnel development carrying international passengers. It covers three different fire and rescue services areas: Kent, Essex and London, with a jointly agreed intervention strategy by all three fire and rescue services.

8C5.6 At an incident the driver will attempt to drive to open air. If that is not achievable the vehicle will stop in an area where the ventilation system will remove the products of the incident, by extraction, or provide cooling by means of fans.

8C5.7 Smoke or fumes will be blown forward of the vehicle to protect any on coming vehicles that may be trapped behind the affected vehicle. The non-affected bore will be positively pressurised to keep it free from contamination.

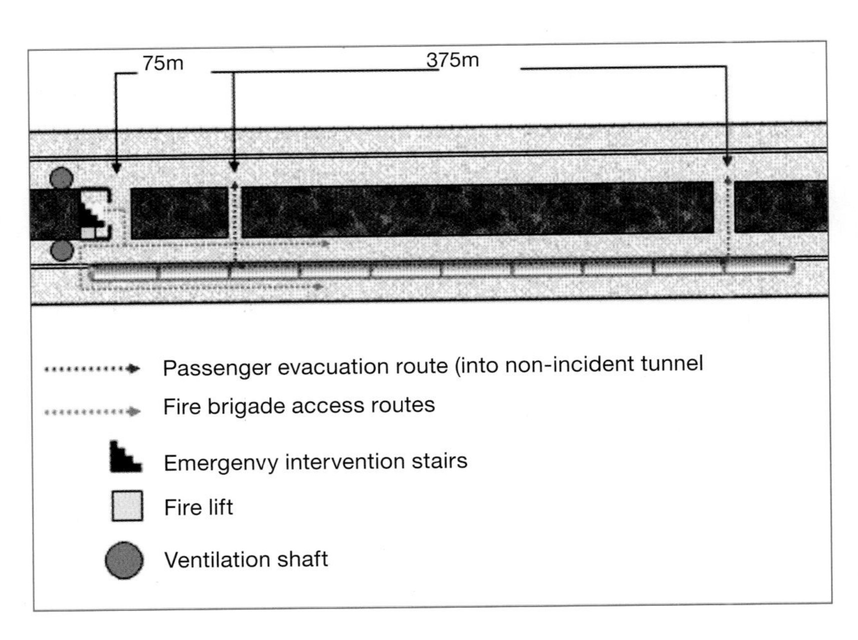

8C5.8 The passengers will be evacuated to the non-affected bore through cross passages. Here they will await an evacuation train to take them to safety The main Fire and Rescue Service response will be via a pressurised shaft. From here rescue and fire fighting operations will commence, using the range of equipment provided for first responders. A secondary attendance will respond to the next nearest shaft, on the other side of the rail vehicle involved.

8C5.9 If the circumstances of the incident are such that this approach is not viable, the use of cross passages, ventilation systems and the unaffected bores will provide alternative options for effective intervention.

8C5.10 When formulating intervention strategies it is crucial to ensure that suitable and sufficient provision is made for the transportation of Fire and Rescue Service staff and equipment into the non incident bore to enable a sustained intervention to be made. In making such arrangements with contractors or infrastructure operators, it is important to establish that Fire and Rescue Service staff must be transported on purpose built and Health and Safety Executive certified transportation. Such arrangements must be included within the intervention planning document

Generic twin bore road tunnel

8C5.11 At a road tunnel the vehicles in the affected bore will drive out, if possible, and vehicles coming into the affected bore will be stopped.

8C5.12 People will use cross passage doors to move to the unaffected bore. Any ventilation system will where possible push the fumes and smoke in the direction of travel to protect those behind the incident.

8C5.13 Responders will attend the incident using both bores with normal traffic flow. This will help to overcome any traffic congestion. Responders entering the affected bore will reconnoitre to identify the location of the incident, looking to obtain protection from the tunnel fire ventilation system. Responders entering the unaffected bore will be protected by the separation and will enter the affected bore using the cross passage.

8C5.14 The control of traffic is essential and will normally be the responsibility of the infrastructure manager.

Generic single bore road tunnel

8C5.15 Here a single lane tunnel attracts an attendance to either portal.

- The Incident Commander gathers information on the best portal to use, depending on the prevailing circumstances of the incident. This will include:
- consideration of the type of incident, the involvement of people and the nature of the infrastructure
- the level of control over smoke or fumes and if any ventilation control system is provided or operating
- breathing apparatus – What type and capacity is necessary for Fire and Rescue Service duties
- communication systems available
- firefighting personnel- The suitability (training, knowledge, experience) and fitness of responders and the likely turn over of crews
- water supplies:
  - how much water will be required to extinguish the fire
  - the different types of fuels and hazards involved
  - the suitability and sufficiency of the water supply
  - the safest method of applying the firefighting media.
- tactical positioning – the position of first and later arriving resources and the position of any cooling/smoke controlling sprays and jets
- communication facilities provided and liaison between Fire and Rescue Service responders.

Generic older infrastructure

5.16 At incidents involving older infrastructure, Incident Commander should consider the following generic points when formulating operational plans:

- it is likely that any means of controlling smoke or fumes will be limited or non-existent

- Fire and Rescue Service attendances will often be made to the nearest Intervention Points on either side of the reported incident
- information should be obtained by the Incident Commander from both attendances, the infrastructure managers
- crews may be committed to undertake specific actions or further reconnaissance depending on the assessment of risk
- crews should be equipped with a suitable level of protection and equipment
- consideration should be given to further crews being committed to secure colleagues' means of escape, and to relay radio transmissions if communications fail
- any action will be largely dependent on the use of the infrastructure, the type of incident and the associated risks to the public, responders and infrastructure.

Section 9

Appendices

APPENDIX A

Tunnels under construction or refurbishment

Introduction

A.1 Construction and refurbishment of tunnels can potentially present increased risks for workers and emergency responders.

A.2 During large tunnelling construction or redevelopment works the risks can change from week to week with progress through the different project stages. It is likely that the facilities responders would normally expect to find in tunnel or underground infrastructure may be absent or degraded, this may include:

- facilities providing fire or smoke separation, including compartmentation
- facilities provided for access to or rescue from the infrastructure, including hard standing, fire fighting stairs, lifts, walkways
- communication facilities to responders or infrastructure managers
- large workforce, unfamiliar with their location or best method of escape
- workers travelling from one contractors area of authority to another's, who may be unfamiliar with the immediate risks in the area
- transient or temporary risks, such as vehicle movements, hot cutting, hazmat transport and storage, structural stability issues.

A.3 Regulations applicable to tunnel construction, require the appointment of a planning supervisor and principal contractor who will be clear contact points for any Fire and rescue Service planning work.

A.4 The role of the Fire and Rescue Service will be to consult with the client and principal contractor to agree the minimum expectations to secure realistic safe systems of work for Fire and Rescue Service responders. This may include developing a statement of expectation, which sets out the facilities to be available to first responders. In securing this understanding the Fire and Rescue Service will indicate, the minimum level of facilities to be in place for the initial Incident Commander to base the first risk assessment upon.

A.5 Facilities to be provided may include the following:

- agreed methods of summoning the Fire and Rescue Service and the information to be provided
- means of fire vehicle access to the entrance of the tunnel or shaft

- means of gathering information on the current layout and risks associated with the development
- secured firefighting water supplies at the surface and in the tunnel
- means of communicating with crews deployed into the tunnel
- agree on access and egress arrangements for crews and evacuation of casualties
- means of communicating with the responsible person at silver.
- facilities to assist with command and control.

A.6 As far as possible the Fire and Rescue Service should promote a standardisation of emergency management arrangements across the entire undertaking to standardise both the Project's and the Fire and Rescue Service's response. For tunnels under construction Fire and Rescue Service personnel should also be aware of BS 6164 Code of Practice for safety in tunnelling in the construction industry.

Construction methods and additional risks

Tunnel boring machines

A.7 These highly specialised vehicles are usually around 120 metres in length, but exceptionally can be up to 250 metres and may be located several kilometres inside a 'dead-end' tunnel. These vehicles would be driven by small trained teams. Tunnel boring machines will normally be used for significant tunnelling projects, and are a large concentration of electrical and hydraulic equipment. In recent years attention has been given to ensuring the level of combustible material is kept to a minimum, by using high flashpoint hydraulic oil, lubricants and greases and by carefully designing the hydraulic hoses to avoid sprays or mists from leaks. Fire suppression systems and good access routes are also essential but there will be residual risks due to certain equipment such as oxygen cylinders, any cutting and burning operations going on, and electrical flashes. Additional support equipment in tunnels under construction may include:

- locomotives train vehicles, normally diesel
- haulage vehicles
- conveyor belts
- ventilation equipment
- high voltage electrical cables
- hazardous materials used during tunnelling process.

A.8 Tunnel boring machines operate by forcing their way forward digging away at the soil, placing concrete rings into place during its progress. The concrete rings will form the basic structure of the tunnel, sealing the joints using specialised grouting compounds. At the same time the tunnel boring machine conveys the spoil to vehicles or conveyor systems for removal.

A.9 The area above the cutting head is the most unstable part of the tunnel structure this area may be pressurised to hold back water and allow workers entry to service the head. Fire and Rescue Service personnel would not enter this area.

A.10 These vehicles are normally carefully manufactured to reduce the combustibles involved to a minimum. They may also incorporate an evacuation chamber that is fire hardened to resist the effects of a fire, providing the workers with a safe haven in the event of a fire. This safe haven may also include communications to the infrastructure manager and an oxygen supply.

A.11 Where an evacuation chamber is not fitted the tunnellers may be instructed to move to a defined safe area, for example a remote chamber, or they may be provided with 're-breather' respirator equipment, to enable them to walk to safety.

A.12 Tunnel boring machines tend to operate with a water main that runs alongside the progress of the machine. Good practice has incorporated conversion of this main into a fire main as the tunnel progresses. This will involve replacing the tunnelling hose connections with standard instantaneous couplings. Some tunnelling boring machines may also be fitted with water curtain facilities to assist with safe egress of workers.

Traditional tunnelling methods

A.13 Tunnellers may use traditional mining methods for parts of the infrastructure. This is particularly the case when forming cross passages, sometimes shored up using wooden supports.

A.14 For hard rock strata tunnellers may use explosives in order to blast an excavation.

A.15 During any planning consultation with those undertaking such work the Fire and Rescue Service should emphasise that there may be limitations on the response that may be provided. This can be affected by, the following:

- the nature and seriousness of any potential incidents.
- the facilities provided to assist the Fire and Rescue Service in their duties
- the quality of information provided to first responders
- the quality of contractors representatives in liaising with first responders
- the effectiveness of the on site emergency plan.

Linings

PRECAST CONCRETE LININGS

A.16 These are constructed in segments, which come together into a ring of tunnel lining.

A.17 The rings may be bolted between segments and bolted between rings or expanded against the ground, and placed adjacent with no connection between the rings. Where rings are expanded there will be certain wedge shaped segment or segments, which are pushed home by one of the rams to expand the ring against the ground. This type of lining is used on the United Kingdom side of the Channel Tunnel.

TUNNELS UTILISING SPRAYED CONCRETE LININGS (SCL) FOR PRIMARY SUPPORT

A.18 The technique of excavating a tunnel and then forming a lining from sprayed concrete is also called the 'New Austrian Tunnelling Method' (NATM). Following a serious collapse in a United Kingdom tunnel under construction the method has been modified and is now referred to as Sprayed concrete lining.

A.19 Most sprayed concrete linings tunnels built in the United Kingdom to date have used the sprayed lining as a primary lining against the ground, and then after erecting a water proof membrane, a secondary lining has been cast against the membrane producing a plain concrete finish. These linings were often reinforced with bar and mesh and so could suffer from explosive spalling problems in a fire.

A.20 More recently, designs such as Hindhead Tunnel use this as a non-structural sacrificial lining for aesthetics and fire protection

CAST IRON LININGS

A.21 These are more correctly currently called spherodial graphite linings. The older linings found on London Underground were made of cast iron which is strong but brittle.

A.22 Today's spherodial graphite linings are more ductile and less brittle but with the same strength. Some of the older tunnels were built entirely with cast iron because it was easily handled and built manually.

A.23 There are many examples of this on the London Underground. Today spherodial graphite linings are only used for specialised areas, such as cross passages, escalators, perhaps station enlargements and where their more slender profile and man handle-ability are required.

STRUCTURAL BRICK LININGS

A.24 Many of the older tunnels used structural brick linings; these are usually a number of layers thick and most are still in sound condition however some do show deterioration over time. They are mostly found in sewerage, canal and water tunnels

CONCRETE

A.25 Following a number of significant fires in tunnels it has been recognised, that in the extreme conditions created by fire in tunnel infrastructure, traditional concrete can spall significantly with a potential for failure of the structural lining. Consequently concrete has been developed which includes 'polyfibres'. These tiny fibres made of, for example polypropylene, are added to the concrete mix. When the set concrete is subjected to heat the fibres evaporate, leaving many tiny cavities. These cavities allow moisture to escape from the heated concrete. This prevents the level of spalling that occurs in conventional concrete.

Additional risks during construction

A.26 Many of the risks related to the construction of tunnels are the same as any heavy civil construction operation, although some are unique, and are exaggerated as a result of being underground. All of the following should be considered during any consultation or planning process, to ensure inclusion in both the Fire and Rescue Service and the contractor's emergency plan.

INUNDATION

A.27 Inundation risk during construction is restricted to bored tunnels and immersed tube tunnels, and needs an understanding of the tunnel profile at any one stage and the potential for entrapment by flooding.

A.28 Inundation/flooding can occur from external sources (the ground, the sea, the river), but also a fracture of pipes in the tunnel carrying slurry, drainage water or cooling water.

ELECTRICAL SHOCK

A.29 The electrical distribution system in a long tunnel under construction may be complex and as the installed power on a tunnel boring machine could be 2-3 megawatts, requires a high voltage feed taken along the tunnel and connected by trailing cables to the back of the machine.

A.30 There will need to be a series of safety trips along the supply line and it is important that these are set properly if they are to be effective. Proper procedures and competent electrical personnel are essential for safety operation and response in an emergency situation.

COLLISION

A.31 There is often not a lot of space in a tunnel under construction; there may be a construction railway with 25 tonne rail vehicles and 120 tonne trains. There may be conveyors and dump trucks. There should be good separation but this can change in an emergency situation. In smaller tunnels, there is significant potential for colliding with equipment and fixtures.

A.32 Many tunnels under construction are accessed by shafts and if there is not a hoist available access will be via a fixed ladder, stairs tower, or man rider skip on a crane.

A.33 At the bottom of a shaft there is always potential for falling objects.

ASPHYXIATION/LACK OF OXYGEN/NOXIOUS GASES

A.34 Noxious gases can be introduced during tunnelling operations. Contractors will normally have good safety arrangements in place to protect workers through air monitoring equipment, however these are not intended for operational personnel.

HEALTH IMPACT

A.35 The health impact to emergency service personnel for tunnels under construction centre on chemicals present/being used for the construction. Certain grouts and grout/cement additives can have health impacts and, will need to be removed from the skin rapidly where contact is made.

PRESSURISED WORKINGS

A.36 During the construction, or the redevelopment, of tunnels or sub-surface infrastructure, compressed air may be used to manage groundwater and stabilise the face of the excavation. Its use must follow strictly controlled procedures both in terms of the impacts on the surrounding ground and impacts on the health of persons working in higher pressures, who like divers can experience the bends and other longer term impacts.

A.37 There are specific regulations and guidance documents and BS6164 gives expanded details.

A.38 It is not envisaged that emergency services personnel would enter a compressed air environment but it is important that they understand what is involved, what the specialist equipment does and the risks and hazards involved.

A.39 Particular to note is that combustion is accelerated in compressed air so that great care is needed to avoid all sources of combustion and processes which could give rise to combustion.

A.40 The other issue is the use of oxygen for decompression which can add to the fire loading at the face of a tunnel boring machine and needs proper care, transport and maintenance procedures.

Emergency management Infrastructure

Tunnel safety management

GENERAL

A.41 Tunnel construction and refurbishment can be an extremely hazardous process, therefore stringent safety management controls must be in place. A detailed risk analysis will have taken place at the planning stage, to determine the management and emergency planning controls that will need to be put in place before construction starts. This will take into account the:

- construction method
- the potential accident/incident causes
- the measures necessary for accident prevention and mitigation and the management controls that need to be applied.

A.42 Fire and Rescue Services will need to consider the impact of the construction geography on Fire and Rescue Service operations for example, the extent of dead end conditions, and any restrictions this may place on intervention distances for firefighting purposes. These limitations should be reflected in discussions between contractor and Fire and Rescue Service and documented in temporary intervention strategies.

Under construction

A.43 During construction entry and exit from a tunnel is normally controlled to allow entry only to personnel who have had appropriate training and induction. The most common control measure is by a tally system; however more complex systems can now be employed such as biometric recognition. Such systems will assist Incident Commanders to ascertain the approximate location of any missing personnel.

A.44 A tunnel controller or 'tally man' will be available when work is in progress to ensure that the entry control system adhered to. The controller will be based in a control room close to the tunnel entry shaft or portal. The controller will know at all times how many people are underground and what locations and type of work processes are being carried out. Tunnel radio/telephone communications are managed from this control room as well as emergency controls for tunnel systems i.e. ventilation, conveyors etc.

A.45 The tunnel controller will be the contact between those underground and the outside world and is likely to be the person to receive emergency calls from underground and have the responsibility for calling for assistance in the event of an accident or emergency.

Under refurbishment

A.46 Safety management requirements for a tunnel under refurbishment will be dependent on the type/use of the tunnel and the extent of the refurbishment and tunnel availability. To enable the refurbishment to take place this could be facilitated by:

- a full closure of the tunnel for the refurbishment period
- closure of part of the tunnel i.e. one bore so the refurbishment can take place and traffic can still operate in a restricted way the other bore
- night-time closures facilitating refurbishment works to be carried out in non-peak periods.

A.47 During a refurbishment project it is likely that some or all of the tunnel safety systems will be unavailable. It is essential that suitable and sufficient alternative control measures are provided; this is particularly relevant if the tunnel remains operational. For traffic/rail tunnels if communication links are difficult to establish with the existing control facility then a temporary local control and monitoring facility, may need to be provided. The temporary local control centre would be provided with communications and be the focal point for the management of alternative temporary systems, examples of these include the following:

- temporary traffic/train management/control systems including an ability to close the tunnel in an emergency
- lighting
- close circuit television cameras systems
- ventilation
- public address systems
- planned and protected means of escape for workers
- fire fighting mains and hydrants
- fire points including the following:
  - fire /emergency procedures
  - means of raising the alarm
  - fire fighting equipment
  - first aid equipment.

Site emergency management structure

A.48 It is important to establish a robust safety management and incident response structure for both tunnels under construction and those under refurbishment, an example of this which has been extensively used is as follows:

MAIN CONTROLLER (TUNNELS SHIFT LEADER)

A.49 The site main controller is the most senior trained on-site member of the management team responsible for exercising functional control over tunnel resources in the event of an incident or emergency. Key to this role is to act as the coordinator between site staff and the emergency services.

INCIDENT CONTROLLER (FOREMEN)

A.50 The most senior trained supervisor closest to the scene of the incident inside the tunnel or in charge of the road works/structures area, who will take charge of the local situation, supervising fire marshals and reporting to the main controller in the event of an incident or emergency.

FIRE MARSHALS/WARDENS

A.51 Fire marshals – selected from engineers, supervisors and construction staff and will be trained and appointed to all work locations inside the tunnel. Ensuring full and safe evacuation of their area in the event of an emergency, and ensure that all staff and visitors report to the assembly points, taking a roll call of all staff and visitors reporting to their assembly point reporting back to main controller.

TUNNEL CONTROLLER (TALLY PERSON)

A.52 Is the person controlling access and egress from the tunnel and is responsible for communications between underground and the outside world. Should an incident occur, the request for assistance from the tunnel to the controller who would be responsible for calling the emergency services.

APPENDIX B
Firefighting and rescue operations in mines

Introduction

B.1 The purpose of this item is to provide Chief Fire Officers with information and guidance on the implications of the arrangements for fire fighting and rescue in mines. Historically and legislatively the Fire and Rescue Service function is limited to providing support for surface operations which could include 'command and control' and limited operations for immediate intervention for rescue subject to limitations of Fire and Rescue Service equipment and training.

B.2 In recent years investment in the Fire and Rescue Service has increased capability in terms of skills and equipment including the specialised areas of urban search and rescue and line rescue. However the Fire and Rescue Service should consider guidance contained within Statutory Instruments 2007 No. 735: Section 3, when making planning decisions about Fire and Rescue Service involvement in underground operations in mines.

B.3 The following guidance explains where local authority Fire and Rescue Services should not become involved in incidents in mines, and those circumstances in which their involvement might be possible. Fire and Rescue Service involvement is subject to proper arrangements and safeguards being in place. This guidance gives a broad indication of the various matters, which should be considered in the pre-planning and implementing of arrangements for firefighting and rescue in mines.

Her Majesty's Inspectorate of Mines

B.4 Her Majesty's Inspectorate of Mines, led by Her Majesty's Chief Inspector of Mines, is part of the Health and Safety Executive and is responsible for enforcing health and safety legislation in mines. Her Majesty's Inspectorate of Mines, embodied in one operational division based in Sheffield, enforce across the whole of England, Scotland and Wales and advise Health and Safety Executive (Northern Ireland) on mining matters.

Legislation

B.5 There is no specific legislation that requires the Fire and Rescue Service to become involved in fire fighting and search and rescue operations in mines. In most circumstances, particularly in coal mines, the limitations and the unsuitability of Fire and Rescue Service equipment (breathing apparatus of insufficient duration and some light alloy and electrical equipment is not approved for use in flammable atmospheres), experience and training impose unacceptable levels of risk, and for these reasons the Fire and Rescue Service would be legally excluded from taking part in unsupervised underground rescue and recovery operations.

B.6 Enforcement responsibility for prevention and control of fire and explosion at mines falls within the auspices the Health and Safety Executive (HMI Mines).

B.7 Statutory Instrument 1995 No. 2870 The Escape and Rescue from Mines

B.8 Regulations 1995 and Health and Safety in Mines Regulations place specific duties on every mine owner to make effective arrangements suitable for the mine for the rescue of persons from the mine and for carrying out work necessary to secure the health and safety of persons below ground in an emergency situation.

B.9 The principle Health and Safety legislation relating to mines is chiefly contained in the Mines and Quarries Act 1954 and Health and Safety at Work Act 1974 and regulations made under the two acts. Firefighting and Rescue operations in mines are dealt with in the Escape and Rescue from Mines Regulations 1995, the Coal and Other Mines (Fire and Rescue) Regulations 1956 (SI 1956/1768) and the Mines Miscellaneous Health and Safety Provisions Regulations 1995. The statutory obligation on the mine owner to make effective arrangements for escape and rescue apply to all mines, irrespective of their size, type or purpose.

B.10 Proportionally, due to their special hazards, the greatest burden to provide effective arrangements for escape and rescue rests with coal mines. However, the use of electrical equipment and diesel machinery underground, storage of flammable material and in the case of tourist mines, large visitor numbers, non-coal mines, storage mines and tourist mines should also have suitable escape and rescue arrangements in place.

The Role of the Fire Service

B.11 Section 7 (2) (a) of the Fire and Rescue Services Act 2004 requires a fire authority to provide 'services for their area of such a Fire and Rescue Service and such equipment as may be necessary to meet efficiently all normal requirements' Underground coal workings present particular and extreme hazards and are not regarded as 'normal' and the fire service does not generally attend mines incidents. This is further reinforced by Statutory Instruments 2007 No. 735: Section 3.

B.12 However, some Fire and Rescue Services have entered into local agreements with mine owners to provide support for mine incidents. This support is generally limited to surface operations such as:

- provision of lighting
- assistance with casualty handling
- provision of specialised equipment
- assistance with demarcation of hazard areas

- assistance in evacuation of a site
- command and control
- communications
- multi agency liaison.

B.13 This support may on rare occasions include limited underground operations requiring immediate intervention for rescue purposes subject to limitations of Fire and Rescue Service equipment, communications and training.

B.14 Notwithstanding the previous paragraph, there are circumstances, where the Fire and Rescue Service is strongly advised not to become involved in underground mines incident response, as outlined below.

Fire and Rescue Service response strategies

B.15 Mines can be divided into two broad categories, namely coal mines and non-coal mines. The potential for local authority Fire and Rescue Service involvement at mines incidents will in general be dependent on the type of mine.

Coal mines

B.16 The Health and Safety Executive and the Department of Business, Innovation and Skills are clear that the special hazards involved in coal mines mean that the rescue can only be effected in safety by experienced miners who are familiar with the mine and specially trained and equipped for the purpose. The legal responsibility for securing the provision of such services is set out in Regulation 13(1), Escape and Rescue from Mines Regulations 1995, which states "no mine of coal shall be worked unless the owner of the mine is a participant in a mines rescue scheme approved by the Secretary of State". There is only one such Scheme in place, that which is administered and operated by the Mines Rescue Service, on a national basis.

B.17 The firm advice of the Health and Safety Executive and the Chief Fire and Rescue Adviser is that local authority Fire and Rescue Service should not enter underground workings in coalmines and to do so would put lives of firefighters in jeopardy. The only possible exception to this general rule is where local authority Fire and Rescue Service are requested by the Mines Rescue Service and Her Majesty's Inspectorate of Mines have confirmed that it is safe for them to do so, an example would be the rescue of casualties trapped by machinery or within transport vehicles.

Non-coal mines

B.18 There are over a hundred non-coal mines of a miscellaneous nature. They can be mines where minerals other than coal are extracted. They can vary in size, some employing only a few staff, others several dozen. There are notable exceptions, which employ hundreds of personnel and/or have extensive underground complexes. The salt mine at Winsford in Cheshire and the Potash mine at Boulby in Cleveland and various gypsum mines located around the country are such examples. Other mines can also include tourist mines and those used only for storage such as, bonded warehouses, military hardware, and document storage. An example being the Wiltshire mines; formerly stone mines that have been adapted for Ministry of Defence use.

B.19 Some of the larger mines have their own rescue teams whilst some smaller mines might have individual trained mines rescue personnel who when combined with other trained personnel from neighbouring mines will form mines rescue teams or act as guides to rescue teams not familiar with the mine layout. Local authority Fire and Rescue Service intending to become involved in incidents at these types of mine must pay close attention to the guidance notes below.

B.20 Disused or abandoned mines are mines that are no longer in operation or have become uneconomic to work and have been abandoned by their owners and no work is being carried out. In these circumstances, enforcement of, and arrangements for safety and rescue and protection against fire and explosion is not the responsibility of the Health and Safety Executive (Her Majesty's Inspectorate of Mines).

B.21 Where local authority Fire and Rescue Services are called on by a member of the public or other organisation to attend an incident involving any type of mine or mine entry, they are strongly advised to immediately seek the advice of the Health and Safety Executive Mines Inspectorate, the Coal Authority and Mines Rescue Service Ltd. As part of any emergency response plan, local authority Fire and Rescue Services should maintain up to date contact telephone numbers.

Pre-planning

B.22 Fire and Rescue Services should take account of information contained in this guidance when planning and establishing local integrated risk management plans.

B.23 Where the owner of a non coal mine wishes to enter into an arrangement with a local authority Fire and Rescue Service for strategic support to fire fighting or rescue emergency arrangements and the Fire and Rescue Service is able efficiently and safely to respond to such a request, it must develop appropriate intervention strategies.

B.24 In parts of the country remote from conventional mines rescue providers, mine owners may ask the local Fire and Rescue Service to provide underground as well as surface emergency fire and rescue response. This can only be acceptable by prior joint agreement with Health and Safety Executive Mines Inspectorate and the Fire and Rescue Service when the mine layout is relatively small and risks of entering the underground workings can be assessed and controlled within the normal operational capability of the Fire and Rescue Service. Such arrangements would need to be reviewed on a regular basis and could not continue once the mine's layout and complexity exceeded the operational capacity of the Fire and Rescue Service personnel and equipment.

B.25 It is essential that all such pre planned intervention strategies are prepared in consultation with the mine owner, the Health and Safety Executive (HMIM) and where appropriate the mines rescue service or any on site mines rescue corps or personnel having responsibility by agreement with the owner for firefighting and rescue purposes. The strategy should determine the precise support role to be undertaken by the Fire and Rescue Service and how that is to be achieved at an emergency incident.

B.26 Fire and Rescue Service intervention strategies in relation to incidents in mines may include the following:

- methods of raising the alarm and alerting the essential services
- speed and weight of the necessary response
- efficiency of inter-service liaison arrangements
- establishment of controls and communications
- initial reconnaissance arrangements
- firefighting procedures (where appropriate)
- rescue and casualty handing
- types of equipment and clothing safe to be used at the scene
- specialised equipment needed
- range of support functions and tasks
- compatibility of plans of other services
- use of mine resources
- the need for specialist training and inter-service training
- media liaison
- health, safety and welfare of personnel
- any other special arrangements or procedures.

B.27 Where Fire and Rescue Services have operational or disused mines in their area or areas where they may respond to, arrangements should be in place to alert the mines rescue service, specialist teams or individuals upon receipt of a call for assistance. There are particular difficulties in pre-planning for incidents in disused mines. Casualties will usually be persons who, having gained unauthorised access, have fallen or become trapped through collapse of tunnels or entrapment within or misuse of abandoned equipment.

B.28 Under no circumstance should Fire and Rescue Service personnel undertake firefighting procedures in underground coalmines. Any support function to firefighting activity must be through prior agreement and through liaison with the mines rescue service at the incident.

B.29 It is the Fire and Rescue Service's responsibility to respond to and resolve fires, rescues and other incidents involving mining industry infrastructure that is above ground. This may include; surface buildings, coal tips, plant and machinery, and other items, which might be found within the boundary of a mine at ground level.

Training

B.30 Where it is either pre-planned, or reasonable to expect that Fire and Rescue Service personnel may be called upon to undertake operations underground, this will normally involve rescue operations in smaller tourist mines. Appropriate training should be devised for Fire and Rescue Service responders in conjunction with the mine owner and Her Majesty's Inspectorate of Mines. This training should be tested periodically, usually through taking part in multi agency emergency response 'mock' exercises in line with agreed Fire and Rescue Service roles and responsibility, and should include the other relevant agencies such as mines rescue service, specialist teams or individuals as applicable.

Firefighting and rescue equipment

B.31 Where local authority Fire and Rescue Services enter into any local agreement they should be mindful of organisational responsibilities with regard to provision of specialised clothing and equipment to enhance personnel safety and welfare.

B.32 When considering any specialised equipment requirements the Fire and Rescue Service should be mindful of *Escape and Rescue from Mines ACOP Appendix 1 and 2*. Further information will be available from the Health and Safety Executive (Her Majesty's Inspectorate of Mines).

Mines Rescue Service Ltd

B.33 Some unusual civil emergency situations may test the capability and capacity of local authority Fire and Rescue Service equipment, skills and knowledge. In these circumstances, it may be useful for the Fire and Rescue Service to consider accessing specialised mining equipment, personnel or knowledge in an effort to reduce risk to personnel.

B.34 Mines Rescue Service Ltd have access to long-duration Breathing Apparatus, specialised ventilation equipment and professional mining teams with the capability of providing complementary assistance in certain circumstances.

B.35 Mines Rescue Service Ltd have a continuously staffed response number, however Fire and Rescue Services should be aware that the lead time for personnel and equipment can be quite lengthy in certain parts of the country

B.36 Training in specialised emergency civil response work can also be provided. Mines Rescue Service Ltd can be contacted through their website which is referenced elsewhere within this document.

APPENDIX C

Human behaviour in tunnels and underground incidents

C.1 Human behaviour in an emergency, particularly a fire emergency can be unpredictable. The purpose of this appendix is to highlight the operational impacts of some aspects of human behaviour at tunnel and underground incidents to assist when involved in any planning or consultation process or incident response.

C.2 It has been recorded on a number of occasions when individuals have ignored warning signs and safety instructions, perhaps unaware of the severity of the developing situation. In tunnels and other underground infrastructure this risk to safety can be compounded by a lack of general understanding of how quickly the conditions can deteriorate and the speed at which smoke and heat will develop.

C.3 In locations where there are a large number of members of the public and a limited number of staff, there have been incidents where people have been unwilling to leave the perceived security of their immediate surroundings, even though a serious fire is developing in clear view.

C.4 In some tunnel fires, tragedies have occurred when people have remained in their vehicles, returned to their vehicles to collect items, or left their vehicles too late and the speed and intensity of the smoke (on one occasion recorded as travelling at 6 metres per second) has overtaken those attempting to escape.

C.5 In road tunnels, drivers have attempted dangerous manoeuvres in order to get themselves and their vehicles out of the tunnel. Or escaping passengers have left their vehicles and locked them, which has blocked access routes.

C.6 People have evacuated from tunnels, returning in the direction they entered. On at least one recorded example this has led them into a hazardous environment.

C.7 The actions of the underground infrastructure's staff at incidents can vary considerably. There is repeated evidence of teams of employees acting selflessly in order to assist members of the public to a place of safety and providing aid to the injured. There have also been instances where key individual employees have not acted in accordance with their training or procedures or who have entered upon an enterprise of their own.

Section 10

Acknowledgements

The Chief Fire and Rescue Advisor is indebted to all those who have contributed their time and expertise towards the production of this operational guidance manual. There were many consultees and contributors, but the following deserve special mention.

Strategic Management Board

Phil Abraham, Fire Service College

John Barton, Retained Firefighters Union

CFO Chris Griffin, Association of Professional Fire Officers

ACFO Peter Hazeldine, Chief Fire Officers Association

Jim Holland, Network Rail

Bob Ibell, British Tunnelling Society

DCFO John Mills, Chief Fire Officers Association

David Morison, Devolved Administration Scotland

Jenny Morris, Health and Safety Executive

Peter Moss, Fire Brigades Union

Neil Orbell, London Fire Brigade Project Manager

Peter O Reilly, Devolved Administration Northern Ireland

Andrew Sargent, Devolved Administration Wales

Craig Thomson, Fire Officers Association

Chris Waters, Institute of Fire Engineers

Core Delivery Group

Sarah Allen, Avon Fire and Rescue Service

Andy Wood, West Yorkshire Fire and Rescue Service

David Bulbrook, London Fire Brigade

Andy Cashmore, West Midlands Fire and Rescue Service

Simon Dedman, Essex Fire and Rescue Service

Jes Eckford Strathclyde Fire and Rescue Service

Graham Gardner, Northumberland Fire and Rescue Service

Graham Gash, Kent Fire and Rescue Service

Rick Hanratty, Avon Fire and Rescue Service

Iain Hunter Wiltshire Fire and Rescue Service

Marc Hudson West Midlands Fire and Rescue Service

David Lovett, Derbyshire Fire and Rescue Service

Chris Noakes Essex Fire and Rescue Service

St John Stanley West Sussex Fire and Rescue Service

Danny Ward, West Midlands Fire and Rescue Service

Industry experts

Tony Aloysius, Highways Agency

Nick Agnew, Transport for London

Andy Barr, Transport for London

Richard Davies, Network Rail

Les Fielding, London Bridge Associates

Tony Forster, HMIM

Neville Hill, HMIM

Jim Holland, Network Rail

Anson Jack, Rail Safety Standards Board

Kevin Lydford, Network Rail

Richard Nowell, Rail Safety Standards Board

Michael Tarran, Heritage Railway Association

David Walmsley, Confederation of Passenger Transport UK

Steve Emery, English Heritage

Organisations

Avon Fire and Rescue Service

British Tunnelling Society

Essex County Fire and Recue Service

Her Majesty's Inspectorate of Mines

London Fire Brigade

Mines Rescue Service Limited

West Midlands Fire and Rescue Service

West Sussex Fire and Rescue Service

Special thanks

Michele Kunneke, Project Support Coordinator

Photographs and illustrations

Thanks to the following individuals, organisations and Fire and Rescue Services for permission to use photographs and diagrams:

Avon Fire and Rescue Service

Greater Manchester Fire and Rescue Service

London Bridge Associates

London Fire Brigade

Network Rail

Thomas Telford publications © images in section 8B 3.8

Transport for London

Schutz and Rettung Feuerwehr and Rettungsdienst (Zurich)

Section 11

Abbreviations and glossary of terms

Although not used throughout the manual as DCLG's house style demands limited use of acronymns we thought it useful to list here those most common in use in the Fire and Rescue Service.

A

AFD – Automatic Fire Detection.

B

BA – Breathing apparatus.

Bi-Directional – Traffic travels in both directions with the same bore.

Bore – A tunnel passage.

British Transport Police (BTP) – a national police force responsible for policing on the majority of rail networks and some tram systems, including the London Underground network.

C

Chief Fire Rescue Advisor (CFRA) – Provides strategic advice and guidance to ministers, civil servants, Fire and Rescue Authorities in England and other stakeholders (including the devolved administrations), on the structure, organisation and performance of the Fire Rescue Service.

Critical National Infrastructure (CNI) – A term used by governments to describe assets that are essential for the functioning of society and economy.

CCA – Civil Contingencies Act.

CCTV – Close circuit television.

CFOA – Chief Fire Officers Association.

D

Dangerous goods – Any product, substance or organism included by its nature of by the regulation in any of the nine United Nations classifications of hazardous materials.

DCOL – Dear Chief Officer Letter.

DCLG – Department of Communities and Local Government.

DFRMO – Defence Fire Risk Management Organisation.

E

EIA – Equality Impact Assessment.

Emergency Response Locations (ERL) – A section of the rail infrastructure where a vehicle is designed to stop in an emergency, and where the facilities are provided to allow firefighting intervention and passenger evacuation.

ERL/IP – Emergency Response Locations/Intervention Points.

F

Fire and Rescue Service (FRS) – Local authority Fire and Rescue Service or local authority Fire and Rescue Authority.

FRA – Fire Rescue Authorities.

FSC – Fire Service Circular.

G

Generic Risk Assessment (GRA) – This is a document that details the assessment of hazards, risks and control measures that relate to any incident attended by the Fire and Rescue Service.

Generic Standard Operating Procedure (GSOP) – Is a list of possible operational actions and possible operational considerations viewed against the 'Managing Incident – Decision Making Model' which has been divided into the six phases of an incident.

Gold/Silver/Bronze Command – The standard management framework employed at complex or major incidents, mandated by the Civic Contingencies Act (2004).

H

H&S – Health and Safety.

HSE – Health and Safety Executive. The government body responsible for protecting people against risks to health and safety arising out of work activities in Great Britain.

Hazard – A hazard is anything that may cause harm.

Hazardous Area Response Team (HART) – Ambulance specialist response teams trained and equipped to work within the inner cordon of an incident.

Hazardous Material (HazMat) – Are referred to as dangerous /hazardous substances or goods, solids, liquids, or gases that can harm people, other living organisms, property, or the environment.

HMIM – Her Majesty's Inspectorate of Mines.

I

ILO – Inter-Agency Liaison Officer.

Infrastructure Managers. (IM) – The Organisation responsible for the management and maintenance of the infrastructure.

Intervention Points (IPs) – A location designated for the use of the Fire and Rescue Service and provided with facilities to ease an emergency intervention

Incident Command System (ICS) – The nationally accepted incident command system, as detailed in *Fire and Rescue Manual Fire Service Operations Volume 2 – Incident Command.*

Incident Commander (IC) – The nominated competent officer having overall responsibility for incident tactical plan and resource management.

Integrated Risk Management Plan (IRMP) – This is the Fire and Rescue Service published assessment of risk within their county/metropolitan boundaries and subsequent action plan to address these risks.

IRU – Incident Response Unit.

IVP/EP – Intervention Point/Evacuation Point.

L

LFB – London Fire Brigade.

Local Authority (LA) – Local government body in a specific area that has the responsibility for providing local facilities and services, e.g. County or District Council.

M

Major incident – A major incidents any emergency that requires the implementation of special arrangements by one or more of the emergency services.

Main Controller (Tunnels Shift Leader) – The Site Main Controller is the most senior trained on site member of the management team responsible for exercising functional control over tunnel resources in the event of an incident or emergency.

MoU – Memorandum of Understanding.

MRSL – Mines Rescue Service Limited.

N

NICS – National Incident Command System.

NPIA – National Policing Improvement Agency.

NATM – New Austrian Tunnelling Method.

O

OLE or OHLE – Overhead Line Equipment, an assembly of metal conductor wires, insulating devices and support structures used to bring a traction supply current to suitably equipped traction units.

P

PDA – Pre-Determined Attendance.

Personal Protective Equipment (PPE) – Provided personal protective equipment issued by the Fire and Rescue Service, includes fire kit, boots, gloves etc.

Piston effect – created by moving traffic pushing the stale air through the tunnel or pushing air through the tunnel to remove stagnant air.

Positive Pressure Ventilation (PPV) – Mobile fan used by the Fire and Rescue Service for ventilating a compartment from smoke and fire gases.

Predetermined Attendance (PDA) – The pre planned Fire and Rescue Service response to accidents/incidents.

Pressurised working – Any work area or worksite that is covered by the Work in Compressed Air Regulations 1996.

PCC Linings – Precast Concrete Linings.

Q

'QDR' – Qualitative Design Review.

R

'Reasonably practicable' – To carry out a duty ' as far as reasonably practicable' means that the degree of risk in a particular activity or environment can be balanced against the time, trouble, cost and physical difficulty of taking measures to avoid risk.

RIDDOR – Reporting of Injuries, Diseases and Dangerous Occurrences Regulations 1995.

Risk – Risk is the probability that somebody could be harmed by a hazard or hazards, together with an indication of how serious the harm could be.

Risk Assessment (RA) – A risk assessment is a careful examination of what, in the workplace, could cause harm to people, in order to weigh up whether enough precautions have been taken or more should be done to prevent harm. The law does not expect the elimination of all risk, but the protection of people as far as is 'reasonably practicable'

Responsible Person at Silver – For Fire and Rescue Service purposes will have the authority, knowledge, training and experience to provide liaison and advice to the IC at 'silver' level, providing information and options to support the overall plan

Respiratory Protective Equipment (RPE) – provided for the protection of Fire and Rescue Service personnel's respiratory system.

RVPs – Rendezvous point. A prearranged meeting place for vehicles and resources attending an incident.

RIDDOR – Reporting of Injuries, Diseases and Dangerous Occurrence Regulations.

S

Safe system of work (SSoW) – A formal procedure resulting from systematic examination of a task to identify all the hazards. Defines safe methods to ensure that hazards are eliminated or risks controlled as far as reasonably practicable.

SCL – Sprayed Concrete Linings

Standard Operating Procedures (SOPs) – Standard methods or rules in which an organisation or Fire and Rescue Service operates to carry out a routine function. Usually these procedures are written in policies and procedures and all firefighters should be well versed in their content.

SCADA – Supervisory Control and Data Acquisition.

SCG – Strategic Co-ordination Group.

SLG – Safety Liaison Group.

Station Control Rooms (SCRs) – Sub Surface stations will be provided with control rooms that will have a range of features and information available to an incident Commander. It is normally accessible via a protected walkway.

T

TBM – Tunnel Boring Machines.

Tactical plan – The operational plan formulated by the incident commander taking into account the objectives to be achieved balanced against identified operational hazards.

TERN – Trans European Road Network.

Thermal imaging camera (TIC) – Type of thermo graphic camera that renders infrared radiation as visible light, it allows firefighters to see and in some cases record temperatures of material.

Tunnel Controller (Tally Man) – The person controlling access and egress from the tunnel and is responsible for communications between underground and the outside world.

U

Urban Search and Rescue (USAR) – Specialist Fire and Rescue Service teams that are equipped to deal with incidents involving the location, extrication, and initial medical stabilisation of casualties trapped in confined spaces.

Uni-Directional – Traffic travels in both directions within the same bore.

W

'Wayfinder' Facilities – various methods for assisting responders and/or tunnel occupants to identify routes.

Section 12

References and supporting information

References and bibliography

HM Government 1974. *Health and Safety at Work etc Act 1974.* London: HMSO

HM Government, 1990. *Environmental Protection Act.* London: HMSO

HM Government, 1991. *Water Resources Act 1991.* London: HMSO

HM Government, 1999. *Management of Health and Safety at Work Regulations 1999.* London: HMSO

HM Government, 2004. *Civil Contingencies Act.* CCA (Contingency Planning) Regulations 2005 London: HMSO

HM Government, 2004. *Fire and Rescue Services Act 2004.* London: HMSO

HM Government, 2007. Fire and Rescue Services (Emergencies) (England), Order 2007. London: HMSO

HM Government, 2008. *Fire and Rescue Manual, Fire Service Operations Vol. 2 Incident Command.* London: TSO

HM Government, 2007. *The Fire and Rescue Services (Emergencies) (England) Order 2007.* London: HMSO

HM Government, 2009. *The Road Tunnel Safety (Amendment) Regulations 2009.* London: HMSO

HM Government, 2007. *The Road Tunnel Safety Regulations 2007.* London: HMSO

HM Government 1998. Manual of Firemanship, Practical Firemanship I Book 11 (8th Impression) Chapter 5.7

HM Government, 1993, *Technical Bulletin 1/1993*. London: HMSO

HM Government, 1997, Statutory Instrument No1713 The Confined Spaces Regulations 1997. London: HMSO

HM Government 1995 Statutory Instrument No. 2870 The Escape and Rescue from Mines Regulations 1995

Directive 2004/54/EC of the European Parliament on minimum safety requirements for tunnels in the trans-European road network.

Fire Research Series 5/2009. *Guide to Risk Assessment tools, techniques and data.* 2009 CLG

Fire Service Circular 55/2004 *The Building Disaster Assessment Group – Key Research Findings* 2004 CLG

2005 *The Handbook of Tunnel Fire Safety* (Alan Beard, Richard Carvel)

Design manual for Roads and Bridges BD78/99 (Highways Agency)

BS 6164, *Code of practice for safety in tunnelling in the construction industry.* 2001 BSI

Safe work in confined spaces Confined spaces regulations 1997, Approved Code of Practice, Regulations and guidance L101 (HSE Books)

Reports/Inquiries

Kings Cross (Fennell, 1987)

Channel Tunnel Fire (HSE 1996)

Baku Underground Metro 1995

Ladbroke Grove (Cullen, 1999)

Manchester technical tunnel fire 2004,

Kingsway Tunnel 2004 (Liverpool)

Limehouse Link Fire 2005.

Fire and Rescue Guidance Documents

Greater Manchester Fire and Rescue Service, November 2009, Incidents Involving Sewers and the Sewer Network, draft operational procedure.

London Fire Brigade, April 1991, Safety in sewers and sewer rescues, Operational Policy 121

London Fire Brigade, July 2007, London Underground (sub-surface) Hazmat procedure and the use of the Motorised Rescue Trolley (MRT), Operational Policy 522

London Fire Brigade, June 2009, Breathing apparatus sub-surface procedure, Operational Policy 467

London Fire Brigade, December 2002, Collapsed structures Operational Policy 467

London Fire Brigade, November 2009, Silos, Operational Policy 681

Merseyside Fire and Rescue Service, July 2009, Tunnels, Standard Operating Procedure 6.1

Web Sites

National

www.communities.gov.uk

www.opsi.gov.uk

www.nao.org.uk

www.networkrail.co.uk/

www.btp.police.uk/

www.hse.gov.uk

www.cabinetoffice.gov.uk

www.tsoshop.co.uk/GRA

www.minesrescue.com

www.highways.gov.uk

www.btplc.com

www.nationalgrid.com

Local

www.channeltunnel.co.uk

www.merseytunnels.co.uk

http://tt2.co.uk/tunnels.php

www.tfl.gov.uk

Section 13

Record of obsolete or superseded previous operational guidance

The table below lists the guidance relating to railway incidents issued by Her Majesty's Government that is now deemed to be obsolete or, is superseded by this Operational Guidance document.

The following abbreviations are used in the table:

- MoF Manual of Firemanship
- DCOL Dear Chief Officer Letter
- FSC Fire Service Circular

Type of guidance	Document title
DCOL 4/1994	Addendum to technical bulletin 1/1993 – operational incidents in tunnels and underground structures
TB 1/1993	Operational incidents in tunnels and underground structures Parts 1 and 2 and App 2
MoF	Manual of Firemanship Book 12 Chapter 5.1 – Sewer Rescues
MoF	Manual of Firemanship Book 12 Chapter 5.2 – Potholes and Mineshafts
DCOL 14/1979	Studies of fire and smoke behaviour relevant to tunnels
DCOL 4/1994	Firefighting and rescue operations in Mines
FSC 17/1971	Explosive Atmospheres In Post Office Cable Chambers

Notes